ESSAYS IN COOPERATIVE GAMES

THEORY AND DECISION LIBRARY

General Editors: W. Leinfellner (*Vienna*) and G. Eberlein (*Munich*)

Series A: Philosophy and Methodology of the Social Sciences

Series B: Mathematical and Statistical Methods

Series C: Game Theory, Mathematical Programming and Operations Research

Series D: System Theory, Knowledge Engineering an Problem Solving

SERIES C: GAME THEORY, MATHEMATICAL PROGRAMMING AND OPERATIONS RESEARCH

VOLUME 36

Scope: Particular attention is paid in this series to game theory and operations research, their formal aspects and their applications to economic, political and social sciences as well as to socio-biology. It will encourage high standards in the application of game-theoretical methods to individual and social decision making.

The titles published in this series are listed at the end of this volume.

ESSAYS IN COOPERATIVE GAMES

In Honor of Guillermo Owen

Edited by

GIANFRANCO GAMBARELLI

University of Bergamo, Italy

KLUWER ACADEMIC PUBLISHERS
BOSTON / DORDRECHT / LONDON

A C.I.P. Catalogue record for this book is available from the Library of Congress.

ISBN 978-1-4419-5260-8 (PB)
ISBN 978-1-4020-2936-3 (e-book)

Published by Kluwer Academic Publishers,
P.O. Box 17, 3300 AA Dordrecht, The Netherlands.

Sold and distributed in North, Central and South America
by Kluwer Academic Publishers,
101 Philip Drive, Norwell, MA 02061, U.S.A.

In all other countries, sold and distributed
by Kluwer Academic Publishers,
P.O. Box 322, 3300 AH Dordrecht, The Netherlands.

Printed on acid-free paper

Table of Contents

1. Foreword vii

2. Acknowledgements viii

3. A Short Biography of Guillermo Owen ix–x

The Dawnings

4. The Coming of Game Theory – *Gianfranco Gambarelli and Guillermo Owen* 1–18

5. Guillermo Owen's Proof of the Minimax Theorem – *Ken Binmore* 19–23

Coalition Formation

6. Encouraging a Coalition Formation – *Michael Maschler* 25–34

Values

7. A Comparison of Non-Transferable Utility Values – *Sergiu Hart* 35–46

8. The *P*-Value for Cost Sharing in Minimum Cost Spanning Tree Situations – *Rodica Branzei, Stefano Moretti, Henk Norde and Stef Tijs* 47–61

9. A Unified Approach to the Myerson Value and the Position Value – *Daniel Gómez, Enrique González-Arangüena, Conrado Manuel, Guillermo Owen and Monica Del Pozo* 63–76

Power Indices

10. α-Decisiveness in Simple Games – *Francesc Carreras* 77–91

11. Monotonicity of Power and Power Measures – *Manfred J. Holler and Stefan Napel* 93–111

12. On the Meaning of Owen–Banzhaf Coalitional Value in Voting Situations – *Annick Laruelle and Federico Valenciano* 113–123

13. "Counting" Power Indices for Games with a Priori Unions – *Marcin Malawski* 125–140

Dynamic Games

14. The Dynamic Stability of Coalitionist Behaviour for Two-Strategy Bimatrix Games – *Ross Cressman, József Garay, Antonino Scarelli and Zoltán Varga* 141–152

15. Dynamic Coalition Formation in the Apex Game – *Emiko Fukuda and Shigeo Muto* 153–163

16. How Many People Can be Controlled in a Group Pursuit Game – *Yaroslavna Pankratova and Svetlana Tarashnina* 165–181

Applications

17. Relevance of Winning Coalitions in Indirect Control of Corporations – *Enrico Denti and Nando Prati* 183–192

18. Takeover Prices and Portfolio Theory – *Gianfranco Gambarelli and Serena Pesce* 193–203

19. A Note on the Owen Set of Linear Programming Games and Nash Equilibria – *Vito Fragnelli* 205–213

20. On the Owen Set of Transportation Situations – *Natividad Llorca, Elisenda Molina, Manuel Pulido and Joaquín Sánchez-Soriano* 215–228

21. The Lovász Extension of Market Games – *Encarnación Algaba, Jesús M. Bilbao, Julio R. Fernández and Andres Jiménez* 229–238

FOREWORD

This special issue of *Theory and Decision* contains a brief history of early Game Theory and collates selected contributions on Game Theory which develop Guillermo Owen's own results. All contributions (except for the first on Minimax Theorem) regard Cooperative Games, of which Owen is undoubtedly a leader. The papers cover both theoretical aspects (Coalition Formation, Values, Simple Games and Dynamic Games) and applied aspects (in Finance, Production, Transportation and Market Games).

These works were written in honour of Guillermo Owen on the occasion of his 65th birthday by authors who participated in the XVth Italian Meeting on Game Theory and Applications, organized by Gian Italo Bischi in Urbino, 9–12 July 2003.

Theory and Decision **56:** vii, 2004.

ACKNOWLEDGEMENTS

I would like to extend my thanks to Guillermo Owen, whose invaluable scientific research formed the basis for this special issue, to Gian Italo Bischi, for his fundamental organizational skills and to all participants at the Meeting who provided the authors with extremely useful comments which helped to improve their contributions. Special thanks to all the experts whose invaluable work in selecting and advice contributed towards making this special issue scientifically sound: Elettra Agliardi, Tamer Basar, Mike Ball, Jesus Mario Bilbao, Steven J. Brams, Dan Fensenthal, Vito Fragnelli, Daniel Gomez, Manfred Holler, Harvo Imai, Rida Laraki, Annick Laruelle, Ehud Lehrer, Moshe Machover, Lina Mallozzi, Jacek Mercik, Dunia Milanesi, Maria Montero, Hannu Nurmi, Francesco Passarelli, Fioravante Patrone, Manuel A. Pulido, Andrea Resti, Joachim Rosenmuller, Agnieszka Rusinowska, Martha Saboyà, Pier Luigi Sacco, Marco Slikker, Stef Tijs, Amparo Urbano, Federico Valenciano, Zoltán Varga, Roy Weintraub, David Wettstein.

I would also like to thank Herma Drees, Arianna Rencurosi and Elena Vorobieva for their technical help, and Marie Sheldon for all her help with the editorial work. Very special thanks go to Cathelijne van Herwaarden for her advice, encouragement and assistance. Finally, I would like to give a heartfelt mention to Ms. Henriëtte Antoinette Beckand Verwée (also known as Nenne).

The Editor

Guillermo Owen

A SHORT BIOGRAPHY OF GUILLERMO OWEN

Guillermo Owen is considered to be one of the "founding fathers" of Game Theory both for his volume *Game Theory* and his contributions to the research at a theoretical and an applicative level. His volume *Game Theory*, published by Academic Press (1967, 1982, 1995) has constituted the "Gutenberg phase" of Game Theory because of its many translations into German, Japanese, Russian, Polish, and Rumanian. His theoretical contributions have been especially expanded on Economic Equilibrium, Multilinear Extensions, Voting and Not-Atomic Games. His applicative contributions have been particularly regarded on Finance, Politics, Economics and studies on Environment, Conflicts and Terrorism.

He was born in Bogotà (Colombia) on 4th May 1938. He moved to Lausanne (Switzerland) where he obtained a diploma at the Lycee Jaccard. In 1957 he moved to Princeton, where he obtained a B.S. a year later. In 1962 he obtained a Ph.D. in Mathematics, with a thesis on Game Theory supervised by Harold Kuhn. His first work experience was carried out at Fordham University (New York), Rice University (Houston, TX) and the University of the Andes (Bogotà). He subsequently started publishing in various scientific reviews, of whose editorial boards he later joined: *International Journal of Game Theory*, *Games and Economic Behaviour*, *Management Science*, *Revista Española de Economia*, and several others. He is now a member of many scientific communities: the (Colombian) Academy of Physical and Exact Sciences, the Royal Academy of Arts and Sciences of Barcelona (Spain) and the Third World Academy of Sciences. He is at present a full professor of mathematics for the U.S. Navy at the Naval Postgraduate School in Monterey (California), where he has been working for the last 20 years and has been Faculty Chair for 10 years.

GIANFRANCO GAMBARELLI and GUILLERMO OWEN

THE COMING OF GAME THEORY

We don't control our lives the way a chess player controls his pieces, but life is not roulette either.

(Ken Follett, 1985, p. 7)

ABSTRACT. This is a brief historical note on game theory. We cover its historical roots (prior to its formal definition in 1944), and look at its development until the late 1960's.

KEY WORDS: Game theory, History

1. INTRODUCTION – THE FOUR PARADIGMS

We would like to think of game theory as the fourth in a line of paradigms, as people learn to optimize results.

Originally, production was unnecessary. In Eden, Adam and Eve received all that they needed without labor. It was only after the fall that man had to earn his bread with the sweat of his brow (Genesis, 3).

1.1. *The First Paradigm*

After the fall, as seen in the book of Genesis, there is a great deal of productive activity. Cain is a farmer; Abel, a shepherd (ch. 4). Noah builds a great ark (ch. 9). Jacob herds sheep for his uncle Laban (ch. 29). Joseph builds great warehouses to store the bounty of the seven fat years (ch. 41).

One thing we do not see in Genesis, however, is any attempt at optimization. Cain and Abel do not use statistical analysis to determine the best crops to plant or the best breeds to tend. Noah

Theory and Decision **56:** 1–18, 2004.

does not attempt to determine an optimal size of ark (he builds according to specifications given him by someone wiser). Joseph does not use sophisticated inventory models for Pharaoh's grain. It could be that the author of Genesis is simply uninterested in the mathematics of optimization; more likely, however, the idea of optimization did not even occur to him and his people.

1.2. *The Second Paradigm*

A new phase arises as people realize that production – indeed, most human activities – can be carried out more or less efficiently. Some mathematics is needed for this: the development of calculus was obviously the most important milestone for optimization, followed closely (in importance) by linear programming, but some attempts at optimization were made even before this. (For example, statistics – numerical summaries of data relating to matters of importance to the state – were collected to determine the most productive activities, as an aid to the state's decision-makers.)

1.3. *The Third Paradigm*

Note that, in optimization, decision-makers generally thought of themselves as working with (or against) a neutral nature. Eventually, however, they realized that their actions influenced the decisions of others who were themselves trying to optimize something – pleasure, utility, etc. Payoffs were a function of both actions. As an example, the demand curve was studied: tradesmen set their prices so as to maximize profits, taking into account both their own costs and potential customers' willingness to pay. As another, police departments choose optimal levels of patrolling, so as to minimize some combination of their own costs and the damages caused by speeding motorists – knowing that motorists' propensity to speed will depend on the level of patrolling.

In this third paradigm, it is necessary to study other individuals' desires (utility). For example, polling data is of great importance for marketing purposes.

Note, however, that the decision-maker does not (in this paradigm) consider the possibility of strategic action by the others. Thus, the monopolist does not consider the possibility that consumers might form a buyers' cooperative to bargain for lower prices or to boycott him. The highway patrol does not consider the possibility that motorists might band together against traffic enforcers, e.g., by signalling the patrolmen's locations to each other.

1.4. *The Fourth Paradigm*

Finally, man realizes that he is interacting, and in competition, with other individuals who are, themselves, aware of this. Thus he must

(a) outsmart others,
(b) learn from others' behaviour,
(c) cooperate with others,
(d) bargain with others.

2. FOREWORD

The modern Game Theory as Interactive Decision-Making is universally recognized as originating in 1944, with the publication of von Neumann and Morgenstern's *The Theory of Games and Economic Behavior*. Earlier studies (even those by von Neumann) had not been introduced in the context of a precise science, which the publication of the above book created. Nevertheless, there were indeed earlier studies. We will thus rather arbitrarily look at the foundations of game theory by studying three general periods: Prehistory – anything up to 1900. Ancient history – 1900 to 1944. Early history – 1944 to 1969. Anything since 1970 will be considered recent history, and thus will not be treated here (though we reserve the right to continue with a subsequent article).

2.1. *Prehistory*

The *minimax* problem (i.e., the existence of equilibrium strategies minimizing the maximum expected loss for each player) dates back to the early 18th Century. At that time, the concept was used by James Waldegrave and Pierre-Rémond de Montmort to analyze the card game *Le Her* (see (Baumol and Goldfeld, 1968)). Waldegrave noticed that the game could essentially be reduced to a 2×2 matrix (all other strategies were dominated), noticed a lack of equilibrium here, and talked about the probability that either player would make the correct guess, but did not actually give a solution in terms of mixed strategies. Later Nicolas and Bernoulli, while studying Le Her, introduced the concept of expected utility and demonstrated its potential applications in Economics (1738). Still in the 18th Century, a preliminary to cooperative game theory appeared in the form of the first known power index, which was developed by Luther Martin in the 1780's (see Riker (1986) for this).

In the 19th Century, we see the general idea of equilibrium strategies emerging, especially in the work of Cournot (1838), who studied the equilibria of an oligopolistic production game. Only pure strategies were considered, but because of the form of the profit function, such equilibria existed in many cases.

Edgeworth (1881) gave a strong impetus to the development of bargaining theory with his introduction of the contract curve. Essentially, he was introducing, in a two-person context, the two important ideas of individual rationality and Pareto-optimality.

In a less academic setting, the general use of interactive strategies can be found in the work of Poe (1845) and Doyle (1891). These two giants of literature both considered a duel of wits between the protagonist (a detective) and an intelligent antagonist, though they lacked the concept of mixed strategies as solutions to the games they considered.

2.2. *Ancient History*

As the 19th Century turned into the 20th, parlor games gave further impulse to the theory. Bertrand (1924), for example, in 1899 studied a simplified form of baccarat in which he emphasized psychological as well as mathematical aspects of the problem. Zermelo proved in 1913 that finite two-person zero-sum games of perfect information (such as chess, checkers, etc.) are strictly determined. Steinhaus studied in 1925 a class of problems, ranging from card games to limited-move chess games, to pursuit. Although significant results were not achieved (due to the considerable combinatorial analysis and differential calculus problems encountered), the work on pursuit was an important precursor of dynamic games.

In the early 20th Century, new concepts were developed. Émile Borel is credited with the first modern formulation of a mixed strategy, in 1921, and with finding the minimax solution for some classes of two-person games in 1924 and in 1927. However, the correct formulation and proof of the general theorem is attributed to von Neumann in his work of (1928). A simpler proof was developed by von Neumann in (1937) using the Brouwer fixed point theorem. Later proofs were based on separation theorems. Among the first of these was one by Borel's student, Ville, in 1938; another was by Weyl, in 1950, with a precise method based on one of his previous works on convex polyhedra (1935).

Numerous other works could be cited, e.g. some military strategy from past centuries.

2.3. *Early History: Princeton*

In the 1930's and 1940's, Princeton had become a living science museum. Here grand old men met, just as in a gentlemen's club, leaving youth to conquer the world. But the minds of the old scientists resonated and the fundamentals that would rule our century were born: new theories in physics, computer science, mathematics, economics. In this greenhouse, ideas germinated, exploiting the synergies of experts from different fields, and from this, modern Game Theory emerged. *The Theory of Games and Economic Behavior* was in fact the collaboration of the mathematician John von Neumann and the economist Morgenstern (1944). Morgenstern recalled nostalgically in 1976 how they occasionally spent evenings together in the company of men such as Einstein, Bohr, and Weyl, and how the ideas for their book matured.

3. ORIGINS AND FIRST APPLICATIONS TO OTHER SCIENCES

The real beginnings of Modern Game Theory can be dated from 1944 for two reasons: first, previous works were fragmentary and lacked organization; second, these works did not attract much attention. With the publication of von Neumann and Morgenstern's book, the Theory of Games had its own concrete organization of fundamental topics at both competitive and cooperative levels. Furthermore, the reputation of the two authors attracted the attention of both mathematicians and economists.

It is fashionable nowadays to say that 1944's *The Theory of Games and Economic Behavior* was merely a recompilation of previous work and contained no new results. In fact, while no new, profound theorems are proved in this book, there are some very important developments. First, the introduction of information sets led to a formal definition of strategy, and thus allowed for the reduction from the extensive to the normal form of the game. In dealing with cooperative games, the treatment of coalitions introduced the characteristic form (admittedly a misnomer) and gave a formal definition to the very important concept of imputation. The introduction of von Neumann–Morgenstern solutions led down a blind alley, but nevertheless included the idea of domination, which would eventually lead to the important concept of the core. The second edition of TGEB (1947) gave a strict development of utility theory.

Thus game theory can really be said to have started – as a formal science – in 1944. The same is true for the computer, which officially appeared in 1944, although the MARK1 was a perfected version of Howard Aiken's ASCC, which dated back to the previous year but was more or less unknown to the general public, because of military secrecy. Here, too, there are numerous precedents from Hollerith to Pascal and so on, back to the first abacus made from "calculi", i.e., stones aligned on the ground.

The coincidence between the beginnings of computers and modern game theory is not due to pure chance, as the genius of von Neumann was a determining factor in the invention of the computer, too. Perhaps if he had not met the English logician,

Alan Turing (a meeting which is even today shrouded in mystery due to war department secrecy), the first computer would have been developed in Germany by Friedrich Zuse, and the course of history would have been changed.

The links between games and computer science were very close at first, especially since both were used for military purposes. Proof exists of the American use of games in wartime by the Operations Evaluation Group of the U.S. Navy and by the Statistical Research Group of the U.S. Air Force (see (Rees 1980)) and in the post-war period by the Rand Project. Subsequently, games and computer science continued to develop together, giving rise to very important synergies. In fact, the first studies in linear programming were developed by George Dantzig in order to solve problems in two-person game theory (see (Albers and Alexanderson, 1985)). Later, games required the application of general mathematics, and this gave the stimulus for research in other fields: fixed points and convex sets for extending the minimax theorem; duality and combinatorics for mathematical programming related to matrix games; and in general topology, probability, statistics, and theory of sets.

The next section deals with the major contributions of Game Theory to the field of economics, which did not occur until much later. Although there were many such well-founded contributions already in the founders' book, economists were slow to appreciate their importance. We will therefore only examine Morgenstern's principal new contributions by quoting Andrew Schotter (who is to be considered Morgenstern's last student):

> Three major things are accomplished. First, the problem for economic science is shifted from a neoclassical world composed of myriad individual Robinson Crusoes existing in isolation and facing fixed parameters against which to maximize, to one of a society of many individuals, each of whose decisions matters. The problem is not how Robinson Crusoe acts when he is shipwrecked, but rather how he acts once Friday arrives. This change of metaphor was a totally new departure for economics, one not appreciated for many years.
>
> Second, the entire issue of cardinal utility is discussed ... An attempt is made to keep the axioms as close as possible to those needed to prove the existence of an ordinal utility function under certainty ...

> Finally, the entire process of modeling exchange as an n-person cooperative game and searching for a "solution" is described ... While the neoclassical theory of price formation was calculus-based and relied on first-order conditions to define equilibrium, game theory, especially cooperative game theory, relied more on solving systems of inequalities ... While the neoclassical solution would often be included within the set of cooperative solutions ... the theory of games offered new and appealing other solutions to the problem of exchange.
>
> (Andrew Schotter, 1992, p. 98)

Thus, beyond the military applications, which did nothing but hamper the early development of the theory (as seen in (Mirowski, 1991)), early interest came from mathematics (where experts were able to find new problems and new mathematical applications) and from economics (where the new models seemed set to completely revolutionize current theory). This caused controversy between the "classicists" and the modern school of thought (see Theocharis (1983) and Arrow and Intriligator (1981)).

Furthermore, there was an important change in mathematical theory. Existing quantitative models in economics had links to physics (do not forget that applied mathematical principles at that time belonged to the world of physics, and economists interested in quantitative methods naturally relied on them). Even Leon Walras and Vilfredo Pareto had engineering backgrounds. This bond was not, however, without its drawbacks, as Giorgio Szegö claimed:

> About 50 years ago it was finally recognized that economic phenomena had certain characteristics that were totally different from those of the physical world, which made them in certain cases completely unsuitable for descriptions of a mechanical nature. This is because a single economic agent behaves not only according to past and present values of certain variables but, contrary to what happens in the physical world, also according to his or her own (possibly not rationally justifiable) expectations about the future values of these quantities. ... Contrary to the situation in mechanics, no invariant law of conservation or variational principle seems to hold for economic systems.
>
> (Giorgio Szegö, 1982, p. 3)

The new approach by von Neumann and Morgenstern, therefore, opened up new horizons. But, contrary to expectations, the first decade of these new models was difficult to digest for the economists, who were unable at that time to understand them in depth. There was therefore no great revolution in the years following 1944 and the classical economists dismissed this new science as nothing more than a passing fad.

4. THE MATHEMATICAL REVOLUTION OF THE 1950'S AND 1960'S

At the middle of the 20th Century, then, the science of game theory had been founded. It was, however, a science still in its infancy. The minimax theorem told us that optimal mixed strategies existed, but, except for 2×2 games and a few other special cases, no efficient methods existed for their computation. Non-zero-sum games were effectively ignored. For n-person games, only a very strange solution concept existed. Further, von Neumann and Morgenstern had made two very strong assumptions, namely, transferable utility (linear side payments) and complete information (full knowledge of the rules and of other players' utility), which, while useful in simplifying the problems studied, narrowed the scope of the theory. Moreover, for all but very small (extensive form) games, the normal form of the game tended to be enormous. (As an example of this, the normal form for such a simple game as tic-tac-toe, even after all symmetries are used to reduce size, is a matrix with over 1000 rows and columns.)

In the first of a series of volumes of papers, *Contributions to the Theory of Games*, Kuhn and Tucker (1950) listed 14 outstanding problems. These included:

(1) To find a computational technique of general applicability for finite zero- sum two-person games with large numbers of pure strategies …

(7) To establish the existence of a solution (in the sense of von Neumann–Morgenstern) of an arbitrary n-person game …

(10) To ascribe a formal "value" to an arbitrary n-person game ...
(12) To develop a comprehensive theory of games in extensive form with which to analyze the role of information ...
(13) To develop a dynamic theory of games ...
(14) How does one characterize and find the solution of games in which each player wishes to maximize some non-linear utility function of the payoff.

As it happened, the first of these problems was soon solved, with two different approaches. On the one hand, Robinson (1951) proved the convergence of the method of fictitious play. On the other, the close relationship between two-person zero-sum games and linear programming meant that the very powerful simplex method developed by George Dantzig (see, on this matter, Koopmans et al. (1951)) could be used for computation of optimal strategies.

As the 1950's continued, a major change in game theory, principally theoretical in nature, took place. The young Princeton mathematicians played a leading role. Foremost among these were Nash and Shapley, though many others contributed.

Nash wrote in (1950a) and (1951) an in-depth general treatment on the notions of equilibrium for non-cooperative games. In this way he took the theory beyond the limits of constant-sum games (on which his predecessors had concentrated) and generalized both the minimax concept (to non-zero sum games) and the results obtained in (1838) by Cournot (to general n-person games). His work on the minimax concept provided a new interpretation of the problem of players' expectations that led to a far-reaching discussion on *refinements of equilibria* which is no less interesting even today. Nash (1950b) also developed the basic methodology for the analysis of bargaining, and in 1953 a no less important concept for the solution of cooperative games with non-transferable utility.

The core was developed by Donald Gillies in his Ph.D. thesis at the Department of Mathematics (1953). This formed the basis for fundamental developments in economics, being as it was a generalization of Edgeworth's (1881) contract curve for the n-person case.

The value concept was introduced by Shapley (1953a) to solve two fundamental problems that until then had effectively stopped the development of the n-person cooperative models: existence and uniqueness. Up until then there had been fruitless attempts to prove a general existence theorem for von Neumann–Morgenstern stable sets for cooperative n-person games. It would be another 20 years before Lucas's counter-examples of (1968 and 1969). Furthermore, the non-uniqueness of the solution had already been noticed by von Neumann and Morgenstern in (1944). Moreover, the core, the set of undominated imputations, could be empty or contain many imputations. Shapley set quite reasonable axioms and determined that there was a unique function (the value) satisfying these axioms for all n-person cooperative games with transferable utility. In (1954) Shapley, collaborating with Shubik, applied this same value to voting games and showed that it could also be thought of as an index of voting power.

Shapley (1953b) also made an important stride in the development of dynamic games by introducing stochastic games, in which the game passes from position to position according to probability distributions influenced by both players. Shapley proved the existence of a value for these games, though they are formally infinite, and found ways of computing optimal strategies.

Shapley's value article was published in the second volume (1953) of the series started at Princeton in 1950 by Kuhn and Tucker. These two mathematicians not only studied constrained optimization, but also made important contributions to the development of Game Theory. Tucker worked with Dantzig on the development of linear programming for the military application of matrix games. Kuhn (1950) analyzed extensive games and in 1952 proposed the first complete presentation and discussion on the proofs of the minimax theorem in existence at that time. Kuhn was also interested in the applications of convex sets to games, and especially in signaling strategies.

Signaling strategies were introduced for the first time in organic form by Gerald Thompson in 1953, and formed the basis of further studies on games with various types of information, which are still of great interest today.

In 1957, Everett generalized Shapley's results on stochastic games by introducing recursive games (the difference being that a stochastic game will terminate with probability 1, whereas a recursive game has positive probability of continuing without end).

At this point we should mention some of the many other intellectuals who were involved in Game Theory at Princeton in the early 1950s: Richard Bellman, David Gale, John Isbell, Samuel Karlin, John Kemeny, John Mayberry, John McCarthy, Harlan and William Mills, Marvin Minsky, Howard Raiffa, Norman Shapiro, Martin Shubik, Laurie Snell and David Yarmish. Later these were joined by Robert Aumann, Ralph Gomory, William Lucas, John Milnor, and Herbert Scarf.

The early 1950's saw the first textbook on game theory (McKinsey, 1952) (this was still a little complex for non-mathematicians), and the first popular editions appeared: McDonald (1950), Riker (1953) and Williams (1954).

In the meantime, the language used by economists was changing, due in good part to the influence of game theory:

> If one looks back to the 1930s from the present and reads in the major economic journals and examines the major treatises, one is struck by a sense of 'the foreign'. ... If, on the other hand, we read economics journal articles published in the 1950s, we are on comfortable terrain: the land is familiar, the language seems sensible and appropriate. Something happened in the decade of the 1940s; during those years economics was transformed from a 'historical' discipline to a 'mathematical' one.
>
> (Roy Weintraub, 1992, p. 3)

As the 1950's ended, Aumann began to study n-person games without transferable utility. (Nash (1953) had only considered two-person games.) Aumann and Peleg (1960) generalized much of the von Neumann–Morgenstern theory to these games.

In 1962, Bondareva neatly characterized games with a non-empty core, in terms of balanced collections of subsets. More or less simultaneously (and independently), Shapley (1965) obtained essentially the same results. These results were generalized to games without transferable utility by Scarf in 1967, using a clever combinatorial argument due to Lemke (1965).

Aumann also collaborated with Maschler (1964) to develop a new solution concept for n-person games: the bargaining set. Maschler in turn, working first with Davis (1965), and then with Maschler and Peleg (1966), developed the important concept of the kernel. Along these same lines, Schmeidler in (1969) developed the nucleolus.

Aumann and Maschler (1963, see also 1995) had been among the first to study the problem of incomplete information in games (i.e., lack of information as to other players' interests). In a series of three important articles, Harsanyi (1967, 1968a, b) developed a sound theory which is still the basis for work done today.

The important developments made since the 1960's cannot be adequately described in a single paragraph. We will, therefore, give a brief outline.

The theory spread from Princeton to the rest of the United States and experienced its first concrete applications. Luce and Raiffa's book (1957), although in a mathematical vein, was more successful than the previous one by McKinsey (1952), as non-mathematicians had by now understood how important it was to make an effort to further their knowledge of the subject. The same can be said for Schelling's publication in 1960 as well as that of Owen (1968), which had some influence in spreading the theory worldwide as these were translated into numerous languages.

ACKNOWLEDGEMENTS

This paper is a new version of an article published in 1994 in the *European Journal of Business Education* (4, 1, pp. 30–45), which appeared in Italian translation in the book '*Giochi competitivi e cooperativi*' (I ed., Cedam, Padova, 1997; II ed., Giappichelli, Torino, 2003). The authors are most grateful to Dunia Milanesi for her invaluable help in updating this paper.

REFERENCES

Albers, D.J., and Alexanderson, G.L. (eds) (1985), *Mathematical People*. Boston: Birkhäuser.

Arrow, K.J. and Intriligator, M.D. (eds) (1981), *Handbook of Mathematical Economics*. Amsterdam: North-Holland.

Aumann, R.J. and Maschler M. (1963), A non-zero-sum game related to a Test Ban Treaty, in Applications of Statistical Methodology to Arms Control and Disarmament. Report to the U.S. Arms Control and Disarmament Agency, 273–287.

Aumann, R.J. (1964), The bargaining set for cooperative games, in M. Dresher, L.S. Shapley and Tucker, A.W. (eds), *Advances in Game Theory, Annals of Mathematics Study* 52, 443–477.

Aumann, R.J. (1995), *Repeated Games with Incomplete Information*, M.I.T. Press (with the collaboration of R. Stearns).

Aumann, R.J. and Peleg, B. (1960), Von Neumann–Morgenstern Solution to Cooperative Games without Side Payments, *Bulletin of the American Mathematical Society* 66, 173–179.

Baumol, W.J. and Goldfeld, S.M. (1968), *Precursors in Mathematical Economics: An Anthology*. Reprinted in: Scarce Works in Political Economy Vol. 19. London: London School of Economics.

Bernoulli, D. (1738), Specimen theorie novae de mensura sortis, in Commentarii Academiae Scientiarum Imperialis Petropolitanae Vol. 5, 175–192. Translated by L. Somer in (1953) as 'Exposition of a New Theory on the Measurement of Risk', Econometrica 22, January, 23–36.

Bertrand, J. (1924), Calcul des probabilités, éléments de la théorie des probabilités, 3rd edn. Paris: Gauthier-Villars.

Bondareva, O. (1962), The Core of an n-Person Game, *Vestnik Leningrad University* 13, 141–142.

Borel, E. (1921), La théorie du jeu et les équations intégrales à noyau symétrique gauche, in Comptes Rendus de l'Académie des Sciences, Vol 173, 1304–1308. Translated by L.J. Savage in (1953) as 'The Theory of Play and Integral Equations with Skew–Symmetric Kernels', Econometrica 21 January, 97–100.

Borel, E. (1924), 'Sur les jeux où interviennent le hasard et l'habileté des joueurs', in J. Hermann (ed.), *Théorie des probabilités*, Paris: Librairie Scientifique. Translated by L.J. Savage in (1953) as 'On Games that Involve Chance and the Skill of Players', Econometrica, 21 (January), 101–115.

Borel, E. (1927), Sur les systèmes de formes linéaires à déterminant symétrique gauche et la théorie générale du jeu, in *Comptes Rendus de l' Académie des Sciences*, Vol. 184, 52–53. Translated by L.J. Savage in (1953) as 'On Systems of Linear Forms of Skew Symmetric Determinant and the General Theory of Play', Econometrica, 21 (January), 116–117.

Cournot, A. (1838), *Recherches sur les principes mathématiques de la théorie des richesses*. M. Rivière et Cie, Paris Translated by A.M. Kelly in 1960 as Researches into the Mathematical Principles of Wealth. New York: A.M. Kelly.

Davis, M. and M. Maschler (1965), The Kernel of a Coperative Game, *Naval Research Log. Quarterly* 12, 223–259.

Doyle, A.C. (1891), The Final Solution.

Edgeworth, F.Y. (1881), *Mathematical Psychics*. London: Routledge and Kegan Paul.

Everett, H. (1957), *Recursive Games. Contributions to the Theory of Games* III, (Princeton), 47–78.

Follett, K. (1985), The Modigliani Scandal. New York: Signet.

Gillies, D. 1953, Some Theorems on *n*-Person Games. Ph.D. Thesis, Department of Mathematics, Princeton University.

Harsanyi, J. (1967), Games with incomplete information played by "Bayesian" Players, I: the basic model, *Management Science* 14(3), 159–182.

Harsanyi, J. (1968a), Games with incomplete information played by "Bayesian" Players, II: Bayesian equilibrium points, *Management Science* 14(5) 320–334.

Harsanyi, J. (1968b), 'Games with incomplete information played by "Bayesian" Players, III: the basic probability distribution of the game', *Management Science* 14, 486–502.

Koopmans, T.C. et al. (1951), *Activity Analysis of Production and Allocation*, New York: Wiley.

Kuhn, H.W. (1950), Extensive Games, *Proceedings of the National Academy of Sciences* 36(10), 570–576.

Kuhn, H.W. (1952), Lectures on the Theory of Games, Report of the Logistics Research Project, Office of Naval Research. Princeton University Press, Princeton.

Kuhn, H.W. and Tucker, A.W. (eds) (1950), Contributions to the theory of games, I. *Annals of Mathematical Studies*, Vol. 20, Princeton: Princeton University Press.

Kuhn, H.W. and Tucker, A.W. (eds) (1953), Contributions to the theory of games, II, *Annals of Mathematical Studies*, Vol. 24, Princeton: Princeton University Press.

Lemke, C.E. (1965), Bimatrix equilibrium points and mathematical programming, *Management Science* 11(7), 681–689.

Lucas, W.F. (1968), A game with no solution, *Bulletin of the American Mathematical Society* 74, 237–239.

Lucas, W.F. (1969), The proof that a game may not have a solution, *Transactions of the American Mathematical Society* 136, 219–229.

Luce, R.D. and Raiffa, H. (1957), *Games and Decisions*. New York: McGraw-Hill.

Maschler, M. and Peleg, B. (1966), A characterization, existence proof, and dimension bounds for the Kernel of a game, *Pacific Journal of Mathematics* 47, 289–328.

McDonald, J. (1950), *Strategy in Poker, Business and War*. New York: Norton.

McKinsey, J.C.C. (1952), *Introduction to the Theory of Games*. New York: McGraw-Hill.

Mirowsky, P. (1991), When games grow deadly serious: the military influence on the evolution of game theory, in C.D. Goodwin (ed.), *Economics and National Security*. Durham, NC: Duke University Press.

Morgenstern, O. (1976), The collaboration between Oskar Morgenstern and John von Neumann on the Theory of games, *Journal of Economic Literature* 14(3), 805–816.

Morgenstern, O. and Von Neumann, J. (1944), *The Theory of Games and Economic Behavior*. Princeton: Princeton University Press.

Nash, J. (1950a), Equilibrium points in *n*-person games, *Proceedings of the National Academy of Sciences, USA*, 36(1) 48–49.

Nash, J. (1950b), The bargaining problem, *Econometrica* 18, 155–162.

Nash, J. (1951), Non-cooperative games, *Annals of Mathematics*, 54(2) 286–295.

Nash, J. (1953), Two-Person Cooperative Games, *Econometrica* 21, 128–140.

Owen, G. (1968), Game Theory, I edn. II edn. 1982, III edn, 1993. New York: Academic Press.

Poe, A. (1845), The Purloined Letter.

Rees, M. (1980), The Mathematical Sciences and Word War II, *American Mathematical Monthly* 87(8) 607–621.

Riker, W.H. (1953), *Democracy in the United States*. New York: McMillan.

Riker, W.H. (1986), The first power index, *Social Choice and Welfare* 3, 293–295.

Robinson, J. (1951), An iterative method of solving a game, *Annals of Mathematics* 54, 286–295.

Scarf, H.E. (1967), The Core of an *n*-person game, *Econometrica* 35, 50–69.

Schelling, T. (1960), *The Strategy of Conflict*. Cambridge, MA: Harward University Press.

Schmeidler, D. (1969), The nucleolus of a characteristic function game, *SIAM Journal of Applied Mathematics* 17, 1163–1170.

Schotter, A. (1992), 'Oskar Morgenstern's contribution to the development of the theory of games', in E.R. Weintraub (ed.), *Towards a History of Game Theory', History of Political Economy*, Annual Supplement to Vol. 24, Durham and London: Duke University Press, pp. 95–112.

Shapley, L.S. (1953a), A value for *n*-person games, in H.W. Kuhn and A.W. Tucker (eds), Contributions to the Theory of Games, II, *Annals of Mathematical Studies*, Vol. 20. Princeton: Princeton University Press, pp. 307–317.

Shapley, L.S. (1953b), Stochastic games, *Proceedings of the National Academy of Science USA*, 39, 1095–1100.

Shapley, L.S. (1965), On balanced sets and cores, Rand Memo RM-4601-PR.

Shapley, L.S. and Shubik, M. (1954), A method for evaluating the distributions of power in a committee system, *American Political Science Review* 48, 787–792.

Steinhaus, H. 1925, Paper in Polish, in *Mysl Akademicka Lwow*, (Vol. 1, pp. 13–14). Translated by E. Rzymovski in (1960) as Definitions for a Theory of Games and Pursuit, with an introduction by H. Kuhn, Naval Research Logistics Quarterly 7(2), 105–108.

Szegö, G.P. (ed.) (1982), Mathematical methods for economic analysis: A biased review, *New Quantitative Techniques for Economic Analysis*. New York: Academic Press, 3–17.

Theocharis, R.D. (1983), *Early Developments in Mathematical Economics*, 2nd edn London: McMillan.

Thompson, G.L. (1953), in H.W. Kuhn and A.W. Tucker (eds), Signalling Strategies in N-Person Games and Bridge Signalling.

Ville, J. (1938), Sur la théorie générale des jeux où intervient l'habileté des joueurs, *Applications des Jeux de Hasard* (E. Borel *et al.* eds), 4(2), 105–113.

von Neumann, J. (1928), Zur theorie der Gesellschaftsspiele, *Mathematische Annalen* 100, 295–320. Translated by S. Bargmann in (1959) in R.D. Luce and A.W. Tucker (eds), as On the Theory of Games of Strategy.

von Neumann, J. (1937), 'Über ein Ökonomisches Gleichungssystem und eine Verallgemeinerung des Brouwerschen Fixpunktsatzes', in K. Menger (ed.), *Ergebnisse eines Mathematischen Seminars. Vienna*. Translated by G. Morton in (1945) as A model of general economic equilibrium, *Review of Economic Studies* 13(1), 1–9.

Weintraub, E.R. (ed.) (1992), Towards a History of Game Theory, *History of Political Economy*, Annual Supplement to Vol 24. Durham and London: Duke University Press.

Weyl, H. (1935), Elementare Theorie der konvexen Polyeder, *Commentarii Mathematici Helvetici* 7, 290–306. Translated by H.W. Kuhn (1950) in H.W. Kuhn and A.W. Tucker (eds) as The Elementary Theory of Convex Polyhedra.

Weyl, H. (1950), Elementary Proof of a Minimax Theorem Due to von Neumann, in H.W. Kuhn and A.W. Tucker (eds), Contributions to the

Theory of Games, I, *Annals of Mathematical Studies*, Vol. 20. Princeton: Princeton University Press, 19–25.
Williams, J.D. (1954), *The Compleet Strategyst*. New York: McGraw-Hill.
Zermelo, E. (1913), Über eine Anwendung der Mengenlehre auf die Theorie des Schachspiels, *Proceedings of V International Congress of Mathematicians* 2, 501–504.

Addresses for correspondence: Gianfranco Gambarelli, Department of Mathematics, Statistics, Computer Science and Applications, University of Bergamo, Via dei Caniana 2, Bergamo 24127, Italy. E-mail: gambarex@-unibg.it

Guillermo Owen, Naval Postgraduate School – Mathematics Code MA/ON, Naval Postgraduate School, Monterey 93943, CA, USA. E-mail: gowen@nps.navy.mil

KEN BINMORE

GUILLERMO OWEN'S PROOF OF THE MINIMAX THEOREM

ABSTRACT. The paper distils the essence of Owen's elementary proof of the minimax theorem by using transfinite induction in an abstract setting.

KEY WORDS: the minimax theorem, zero-sum games, game theory

1. INTRODUCTION

This tribute to Owen (1982) looks at his elementary proof of Von Neumann's minimax theorem. I think the elegance of Owen's argument is best appreciated when it is used to prove the theorem in a more abstract setting. The sketch of the argument in my game theory textbook (Binmore, 1992) fails to get this point across, because the transfinite induction I took for granted there turns out to be a less familiar mathematical tool than I had realized.

2. MINIMAX THEOREM

We consider two-person zero-sum games in which the strategy sets P and Q are convex and compact, and the first player's payoff function $\Pi : P \times Q \to \mathbb{R}$ is concave and continuous.

The minimax and maximin values of the game are defined by

$$\underline{v} = \max_{p \in P} \min_{q \in Q} \Pi(p, q) = \min_{q \in Q} \Pi(\tilde{p}, q), \tag{1}$$

$$\bar{v} = \min_{q \in Q} \max_{p \in P} \Pi(p, q) = \max_{p \in P} \Pi(p, \tilde{q}), \tag{2}$$

where $\tilde{p}$ is a strategy p in P for which $\min_{q \in Q} \Pi(p, q)$ is largest, and $\tilde{q}$ is a strategy q in Q for which $\max_{p \in P} \Pi(p, q)$ is smallest.

The first player can guarantee a payoff $\underline{v}$ or more by playing $\tilde{p}$. The second player can guaranteed a payoff of $-\bar{v}$ or more by

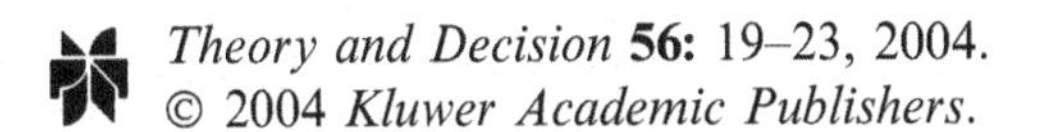
Theory and Decision **56:** 19–23, 2004.

playing $\tilde{q}$. Since the game is zero-sum, it follows that $\underline{v} - \bar{v} \leq 0$. Hence $\underline{v} \leq \bar{v}$.

The pair $(\tilde{p}, \tilde{q})$ is a Nash equilibrium of the game if and only if it is a saddle point of the payoff function Π. It is easy to check that this is true if and only if $\tilde{p}$ and $\tilde{q}$ are given by (1) and (2), and $\underline{v} = \bar{v}$. It is for this reason that Von Neumann was led to prove the most celebrated result in game theory.

2.1. *Minimax Theorem*

$$\underline{v} = \bar{v}.$$

Various proofs of the minimax theorem have been given using fixed point theorems or the theorem of the separating hyperplane, but Owen's proof by induction is entirely elementary. My version of his proof is simpler still, but requires using transfinite ordinals.

3. ORDINALS

The finite ordinals $0, 1, 2, \ldots$ are used to count finite sets. They need to be supplemented with the transfinite ordinals when counting infinite sets. When we have used up all the ordinals we have constructed so far, we invent a new ordinal to count the next member of a well-ordered set. For example, if we run out of finite ordinals when counting an infinite set, we count its next element with the first transfinite ordinal, which is denoted by ω.

The cardinality of a set is often identified with the ordinal that follows the smallest set of ordinals needed to count it. For example, the set $\{a, b, c, d\}$ has four elements because it can be put into one-to-one correspondence with the set $\{0, 1, 2, 3\}$. The cardinality of the set of all rational numbers is ω because it can be put into one-to-one correspondence with the set $\{0, 1, 2, 3, \ldots\}$.

4. PROOF

We show that the assumption $\underline{v} < \bar{v}$ implies a contradiction. The minimax theorem then follows from the fact that $\underline{v} \leq \bar{v}$.

The proof requires the construction of a zero-sum game for each ordinal α which has *nonempty*, convex and compact strategy sets P_α and Q_α, but the same payoff function as the original game. The first of these games is identical with our original game, so that $P_0 \times Q_0 = P \times Q$. Later games get progressively smaller, in the sense that $\alpha < \beta$ implies $P_\beta \times Q_\beta \subset P_\alpha \times Q_\alpha$, where it is important that the inclusion is *strict*.

The reason that this construction yields the desired contradiction is that $P_\gamma \times Q_\gamma$ must be empty if γ is a sufficiently large ordinal, because one cannot discard more points from $P \times Q$ than it contains.

4.1. *The First Step of the Construction*

Owen's ingenious idea is to replace $P \times Q$ by $P_1 \times Q_1$ in a manner that makes

$$\bar{v}_1 - \underline{v}_1 \geq \bar{v} - \underline{v}. \tag{3}$$

If $\underline{v} \geq \Pi(\tilde{p}, \tilde{q})$, and $\Pi(\tilde{p}, \tilde{q}) \geq \bar{v}$, then $\underline{v} \geq \bar{v}$, It follows that our assumption that $\underline{v} < \bar{v}$ implies that either $\underline{v} < \Pi(\tilde{p}, \tilde{q})$ or $\Pi(\tilde{p}, \tilde{q}) < \bar{v}$. The former inequality will be assumed to hold. If the latter inequality holds, a parallel argument is necessary in which it is P that shrinks rather than Q as in the line followed below.

Take Q_1 to be the *nonempty*, convex and compact set consisting of all q in Q for which

$$\Pi(\tilde{p}, q) < \underline{v} + \varepsilon, \tag{4}$$

where $0 < \varepsilon < \Pi(\tilde{p}, \tilde{q}) - \underline{v}$. Then Q_1 is strictly smaller than Q because it does not contain $\tilde{q}$. Let $P_1 = P$.

With $\tilde{p}_1$ and $\tilde{q}_1$ defined in the obvious way, consider the convex combinations $\hat{p} = \alpha\tilde{p} + \beta\tilde{p}_1$ and $\hat{q} = \alpha\tilde{q} + \beta\tilde{q}_1$. Observe that

$$\begin{aligned}
\bar{v} = \min_{q \in Q} \max_{p \in P} \Pi(p, q) &\leq \max_{p \in P}(p, \hat{q}) \\
&= \max_{p \in P}\{\alpha\Pi(p, \tilde{q}) + \beta\Pi(p, \tilde{q}_1)\} \\
&\leq \alpha \max_{p \in P} \Pi(p, \tilde{q}) + \beta \max_{p \in P_1} \Pi(p, \tilde{q}_1) \\
&= \alpha\bar{v} + \beta\bar{v}.
\end{aligned} \tag{5}$$

An inequality for $\underline{v}$ requires more effort. Note to begin with that

$$\begin{aligned}\min_{q\in Q_1}\Pi(\hat{p},q) &\geq \alpha\min_{q\in Q_1}\Pi(\tilde{p},q)+\beta\min_{q\in Q_1}(\tilde{p}_1,q)\\ &\geq \alpha\min_{q\in Q}\Pi(\tilde{p},q)+\beta\min_{q\in Q_1}(\tilde{p}_1,q)\\ &= \alpha\underline{v}+\beta\underline{v}_1.\end{aligned} \tag{6}$$

Also,

$$\begin{aligned}\inf_{q\notin Q_1}\Pi(\hat{p},q) &\geq \alpha\inf_{q\notin Q_1}\Pi(\tilde{p},q)+\beta\inf_{q\notin Q_1}\Pi(\tilde{p}_1,q)\\ &\geq \alpha(\underline{v}+\varepsilon)+\beta c.\end{aligned} \tag{7}$$

At the second step of the preceding derivation, recall that, if $\Pi(\tilde{p},q)\leq \underline{v}+\varepsilon$, then q lies in the set Q_1 by (4). The constant c is simply an abbreviation for $\inf_{q\notin Q_1}\Pi(\tilde{p}_1,q)$.

We want (6) to be smaller than (7). To arrange this, $\alpha=1-\beta$ and β have to be carefully chosen. By taking β to be very small, (6) can be made as close to $\underline{v}$ as we choose. Similarly (7) can be made as close to $\underline{v}+\varepsilon$ as we choose. Thus, if β is chosen to be sufficiently small, then (6) is less than (7). However, it is important that β is not actually zero.

An inequality for $\underline{v}$ now possible:

$$\begin{aligned}\underline{v}=\max_{p\in P}\min_{q\in Q}\Pi(p,q) &\geq \min_{q\in Q}\Pi(\hat{p},q)\\ &= \min\left\{\min_{q\in Q_1}\Pi(\hat{p},q),\ \inf_{q\notin Q_1}\Pi(\hat{p},q)\right\}\\ &\geq \min\{\alpha\underline{v}+\beta\underline{v}_1,\alpha(\underline{v}+\varepsilon)+\beta c\}\\ &= \alpha\underline{v}+\beta\underline{v}_1.\end{aligned} \tag{8}$$

The desired inequality (3) can then be deduced from (8) and (5).

4.2. *Continuing the Construction*

When an ordinal β has an immediate predecessor α, the preceding discussion explains how we construct nonempty, convex and compact subsets P_β and Q_β of P_α and Q_α with

$$\bar{v}_\beta-\underline{v}_\beta\geq\bar{v}_\alpha-\underline{v}_\alpha. \tag{9}$$

When β does not have an immediate predecessor, we simply take P_β to be the intersection of all P_α with $\alpha < \beta$, and Q_β to be the intersection of all Q_α with $\alpha < \beta$. The continuity of the payoff function then assures that (9) holds whenever $\alpha < \beta$. The fact that each P_α and Q_α is nonempty, convex and compact assures that the same is true of P_β and Q_β. It is also true that the inclusion $P_\beta \times Q_\beta \subset P_\alpha \times Q_\alpha$ is strict when $\alpha < \beta$.

This concludes the construction and hence the proof of the minimax theorem.

REFERENCES

Binmore, K. (1992), *Fun and Games*. Lexington, MA: D.C. Heath.
Owen, G. (1982), *Game Theory*, 2nd edn. New York: Academic Press.

Address for correspondence: Ken Binmore, Department of Economics, University College London, Gower Street, London, WC1E 6BT, UK. Tel.: +44-171-504-5881; Fax: +44-171-916-27745; E-mail: uctpa97@ucl.ac.uk

MICHAEL MASCHLER

ENCOURAGING A COALITION FORMATION

ABSTRACT. A 4-person quota game is analyzed and discussed, in which players find it beneficial to pay others, in order to encourage favorable coalition structure.

KEY WORDS: Game theory, cooperative games, power of a coalition, coalition formations, experiments in game theory.

1. INTRODUCTION

In Maschler (1978) I described a set of 123 cooperative games played by high school students. Among the games there were 18 *4-person quota games* with varying quotas $\omega = [\omega_1, \omega_2, \omega_3, \omega_4]$.

Specifically, groups of four students were told who among them, is player $1, \ldots, 4$. If any 2-members i, j decided to form a coalition, they would receive together $\omega_i + \omega_j$ "points", about which they had to decide how to share among themselves and report on a card. The three students who collected the highest number of points by the end of the 123 plays would then get prizes which were coupons for buying books. To boost their ambition I also told them that they should make every effort to gain as many points as possible, because this experiment is scientifically important. Either because of this encouragement, or because of the enjoyment from the competition, they fought fiercely to gain points.

At the end of each play, the participants were asked to report briefly on the back of the card what their reasonings were while playing, and why they agreed to the points they received.

Note that I did not reveal how I determined the characteristic function. I simply provided them with the characteristic function. Nevertheless, the students, who belonged to an elite class, were intelligent enough to realize how the worths of the coalitions were obtained.

Theory and Decision **56:** 25–34, 2004.

In this paper I will use the quota $\omega = [10, 20, 30, 40]$ as a prototype, even though, as said earlier, the experiment used various quotas. For the above quota game, we were interested in finding out if the players would end up anywhere near the nucleolus of the game (10,20,30,40). A deviation of 5 points or less was judged a "success", as this small amount was regarded the least noticeable difference. (10,20,30,40) is the nucleolus of the game, no matter which couple of 2-person coalitions form.[1]

Altogether, 18 plays were performed of which one ended up with the students entering a fight.[2] This game is discarded. Of the 17 games, 9 ended successfully, as predicted. But 8 ended up far from the quota vector.

Nevertheless, even in these games there was always one coalition that ended up within the tolerable distance from the quota. For example, an outcome could be $(22, 23, 27, 28; \{23, 14\})$ where players 2 and 3 ended up near their quota but players 1 and 4 were far off.[3]

This result is interesting as it conforms to a von Neuman Morgenstern solution, discovered by Shapley (1953). Obviously, the students new nothing about von Neumann Morgenstern solutions, or any other game theoretical solutions. What was it that caused them to act so strangely? The answer became obvious after I read the reports of the students and realized that the coalition that reached the neighborhood of the quota was always the coalition that was first to form. Now, everything became transparent: As long as no coalition is formed, there is a pressure to share in accordance with the quota. When two players are left alone, the quota is meaningless. Why not share the proceeds equally, for example? If in the above example, once coalition $\{2, 3\}$ was formed there remained two players, 1 and 4 who together would earn 50 points and alone each was worth nothing. Why not share (25,25)? Indeed, in all the 8 cases, the last coalition shared its proceeds somewhere between the quota share and the equal split share. Why not sharing equally when realizing that they were left alone? Is it because of benevolence, to compensate the player with the larger quota who became disappointed, realizing that suddenly he lost his potential wealth? Is it in order to compensate for his regret for not succeeding to be in the first coalition to form? Perhaps psychologists can explain this. I do want to stress, however, that from the few plays that I watched personally and from the reports, it seems to me that the player with the higher quota would have refused an equal share.

In two plays Player 1 was smart. He realized that as long as all players were negotiating he could expect only an amount near 10, but if he was left out, he would have a good chance to improve his lot. In these cases (one of which I happened to watch when it was played), this player made excessive demands so that nobody was willing to join him. The three players competed frantically in order to enter a 2-person coalition, negotiating a quota share. Once a single person was left to negotiate with Player 1, it was easy for Player 1 to explain that now, since only two were left, they should share equally; however, he is benevolent and is willing to give the other player a token more.

For many years, in private conversations and in class, I explained that this is an example in which high school kids outsmarted game theory. The two students realized that it was smarter to wait and perhaps the others realized that it is in their interest to rush to form a coalition so as not to be stuck with Player 1. There does not exist a solid theory that tells us, in the sense described above, when to rush and form coalitions, when it is advantageous to wait, with whom to join and how much to ask for.[4] I added that with all due respect to game theory, if I were Player 1, I would have acted similarly.

Nowadays I am somewhat smarter than the two students, as will be described in the next section.

2. SOLVING THE GAME, USING THE POWER OF COALITIONS

We analyze the game from a normative point of view. To do so, we assume that the last coalition that forms shares its worth equally among its members. This is the *standard of fairness* that we assume to be prevailing with the players. Although in reality the players deviated from this standard, as explained above, we regard these deviations as being an order of second magnitude and choose to ignore them.

As explained in the introduction, it is beneficial for Player 1 to wait until a 2 person coalition forms. However, Player 1 would like that Players 2 and 3 form a coalition, because, being left with Player 4, he would raise his profits from a quota of 10 utils to 25 utils. Any other 2-person coalition that forms among players 2, 3 and 4 will yield him less. *He might be willing to pay some utils to players 2 and 3 in order to "encourage" the formation of this coalition. How much should he offer them?!*

To answer that, we recall Maschler (1963), where it is argued that when a standard of fairness exists among the players then, even if a game is presented to them by a characteristic function, the players perceive the game as having a different worth function, called, *the power function.*[5] Denoting the power function by w and assuming the above standard of fairness, we proceed as follows:

For 2-person coalitions $\{i,j\}$, $w(ij) = v(ij) = \omega_i + \omega_j$. Indeed, this is all that these coalitions can guarantee. A 3-person coalition $\{i,j,k\}$ can do more than $\max\ \{\omega_\mu + \omega_\nu : \mu, \nu \in \{i,j,k\}, \mu \neq \nu\}$. Indeed, they can decide which 2-person coalition forms and who will remain to form a coalition with the fourth player. Thus, coalition $\{1,2,3\}$ will maximize its proceeds if its members decide to form coalition $\{2,3\}$ and let Player 1 play with Player 4 and get 25 utils. Thus, $w(123) = 50 + 25 = 75$. Similarly, $w(124) = 80$, $w(134) = 85$, $w(234) = 85$. Single-person coalitions can guarantee only what their complementary coalitions spare them. Thus, $w(1) = 15$, because Player 1 can count on the fact that at least Player 2 will join him[6] $w(2) = 15$, $w(3) = 20$, $w(4) = 25$. Finally, $w(\phi) = 0$ and $w(1234) = 100$, as the players can safely assume that eventually a couple of 2-person coalitions will form. Note that the power function is a constant sum game, so, trying to form a "new power" of the power function will not yield any different function.

We propose that the players look for the nucleoli of the game $(\{1,2,3,4\};\ w\})$ for the various coalition structures.[7]

For example, if Player 1 wants that coalition $\{2,3\}$ form, he should look for the nucleolus of the power game for the coalition structure $\{\{1,2,3\},\{4\}\}$ as he is trying to form the 3-person coalition $\{1,2,3\}$ to negotiate a compensation to players 2 and 3. The nucleolus vector for this coalition structure is

$$(17.5, 25, 32.5, 25).$$

This share is quite reasonable! It can be described by the following proposal of Player 1 to $\{2,3\}$: "If you form a coalition then my proceeds will rise buy 15 (from a quota of 10 to 25 utils). I give you half of it, namely 7.5, by offering 5 utils to Player 2 and 2.5 utils to Player 3. The reason for the unequal division is because it is more valuable for Player 1 that Player 2 enters a 2-person coalition with either Player 3 or 4. This will raise his proceed by at least 10 (from 10 to 20), whereas if Player 3 forms a coalition, he may be stuck with Player 2, in which case his proceeds will rise only by 5 (from 10 to 15). Thus in effect, Player 1 offers each of the players 2 and 3 half of the minimal excess amount each guarantees him by entering a coalition before he forms a coalition. I cannot imagine a better resolution of the question asked at the beginning of this section. I did not expect such a convincing resolution before looking at the nucleolus and I am grateful that the nucleolus revealed it.

If Player 1 succeeds to "bribe" 2 and 3 to form a coalition, then Player 4 is the big loser. He might try to "bribe" either Player 2 or Player 3 to form a coalition with him. This is going to cost him. How much?

Heuristically, it seems that he should approach Player 3 and offer him 32.5 utils to outbid the offer of Player 1. This is better than trying to outbid by offering 25 utils to Player 2, thereby losing 15 utils from his quota. Indeed, the nucleolus of the power game for the coalition structure $\{\{1,2\},\{3,4\}\}$ is[8]

$$(15, 15, 32.5, 37.5).$$

Note that the nucleolus drives Player 1 to his power worth.

The nucleoli of the power game for the various coalition structures in which the total amount of proceeds is 100 utils is given below. We write the coalitions that form, as much as possible, in the order of formation that, we believe, justifies the outcome:

$$\left(17\frac{1}{2}, 25, 32\frac{1}{2}, 25; \{123, 4\}\right),$$
$$\left(20, 22\frac{1}{2}, 20, 37\frac{1}{2}; \{124, 3\}\right),$$
$$\left(22\frac{1}{2}, 15, 27\frac{1}{2}, 35; \{134, 2\}\right),$$
$$\left(15, 23\frac{1}{3}, 28\frac{1}{3}, 33\frac{1}{3}; \{234, 1\}\right),$$
$$\left(15, 15, 32\frac{1}{2}, 37\frac{1}{2}; \{34, 12\}\right),$$
$$\left(16\frac{2}{3}, 21\frac{2}{3}, 28\frac{1}{3}, 33\frac{1}{3}; \{23, 14\}\right),$$
$$(15, 25, 25, 35; \{24, 13\}),$$
$$\left(17\frac{1}{2}, 20\frac{5}{6}, 28\frac{1}{3}, 33\frac{1}{3}; \{1234\}\right).$$

Assuming that the players consider the nucleolus as their solution, it seems that player 1 is in quite a strong position. The only way Player 4 can compete is by either convincing Player 3 or Player 2 to join him in a coalition that forms first. All other coalition structures yield both players less than the offer in the coalition structure $\{123, 4\}$. If Player 4 succeeds to attract either Player 3 or Player 4, Player 1 is driven to his bottom worth of 15.[9]

We are not claiming that either $\{123, 4\}$ or $\{34, 12\}$ will form. An ambitious Player 4, for example, may opt for $\{124, 3\}$ or $\{34, 12\}$ hoping for 37.5 utils, thereby attempting to cause a competition between the losers Player 2 or Player 3 to gain his favor. The formation of the grand coalition is also interesting: It can happen when Player 1 fears that either Player 2 or Player 3 will join Player 4 who has his own fears. Then, players 1 and 4 may agree to cause the formation of the grand coalition by committing themselves not to listen to offers from Player 2 and/or Player 3. Player 2 can then try to destroy this coalition by forming a coalition with Player 3, but this may not succeed, especially if players 1 and 4 promise an additional token to Player 3. What we want to argue is that when the order of formation of coalitions is considered, and the players regard the nucleolus as their solution, then there are interesting bids and counter bids on various coalition formations.

3. CONCLUSION

We started with a game with a solution that looked trivial: A couple of 2-person coalitions will form and no-matter what, the payoff will be in accordance with the quota. This argument does not take into account that coalitions do not form simultaneously. It also does not take into account that a standard of fairness exists, according to which (at least normatively), *if all things are equal let us share equally*. Our analysis revealed that coalition formation is relevant and drastically affects the payoff outcome. One still cannot predict which coalitions will form, but whatever structure forms determines the payoff vector. The coalitions that form and the order of formation are strategic variables, and in order to achieve a desirable formation one may have to pay the participants. The present analysis suggests that the nucleolus of the power game can be employed to decide how much each player should get theoretically. For example, one cannot predict in our example if the coalition structure will be e.g.[10], $\{\{1,2,3\},\{4\}\}$ or $\{\{3,4\},\{1,2\}\}$, if Player 1 takes the initiative and tries to convince Players 2 and 3 to form a coalition. Presumably, the final actual outcome will be determined by who will be willing to sacrifice somewhat more from the theoretical nucleolus prediction as in the spirit of the bargaining set theory. The theoretical nucleolus of the power game for a coalition structure predicts only a "center" around which real outcomes in games played by knowledgeable players should come about.[11]

We have based our analysis on a single principle; namely, that a standard of fairness exists in the population of the players. Can one generalize this analysis to a general solution theory for TU cooperative games? I feel that such task is not easy to achieve. Issues that should be addressed are: what types of standards of fairness can one assume when a subgame based on a certain coalition is not a unanimity game? How then should a power function be defined? More importantly, issues concerning coalition formation and the order of the formation can be more involved: a player may want that coalition S form provided that coalition T *does not form*; otherwise he would prefer the formation of another coalition, etc. It is therefore quite a challenge to generalize these ideas to a general theory.

ACKNOWLEDGEMENTS

I wish to express my gratitude to many game theorists with whom I discussed this example and in particular to Bezalel Peleg with whom I spent many hours discussing the general issues that stem out of this example.

NOTES

1. Regardless of whether we attribute 0 to the worth of the 3 and 4-person coalitions or take the super additive cover of this game.
2. A student reported that he caused a rift between players 1 and 4.
3. Whenever we find it convenient, we shall refrain from using commas and curly brackets when specifying a coalition. Thus, we often write 12, instead of $\{1,2\}$.
4. Of course there are several papers concerning coalition formation that look at the issue in other senses.
5. Rapoport and Kahan investigated experimentally the existence of standard of fairness and power perceptions among players playing characteristic function games. See Rapoport and Kahan (1979, 1982) and Kahan and Rapoport (1984), which summarizes their findings.
6. We take into consideration that Player 1 can wait until a 2-person coalition forms. We could be less biased and more pessimistic, allowing for the possibility that he may be forced to be in the coalition that forms first. We will then have to define $w(1) = \min\{10, 15\} = 10$. It turns out that the insight gained in this paper does not change much.
7. For the definition and computation of the nucleolus for a game when a coalition structure forms, see Schmeidler (1969), Owen (1977), Wall-

meier (1980), Potters and Tijs (1992), Maschler et al. (1992). In our game the easiest way to compute the nucleolus when a three person coalition forms, is to compute its kernel which consists of a single imputation and therefore equals the nucleolus (see Maschler and Peleg, 1966). When a couple of 2-person coalitions form, perhaps brute force is the easiest way to compute the nucleolus.

8. It is (11.25, 18.75, 32.5, 37.5) if we decide that $w(1) = 10$.
9. If we define $w(1) = 10$, three imputations change as follows: $(10, 23\frac{1}{3}, 28\frac{1}{3}, 33\frac{1}{3};\ \{234, 1\})$, $(11\frac{1}{4}, 18\frac{3}{4}, 32\frac{1}{2}, 37\frac{1}{2};\ \{34, 12\})$, $(12\frac{1}{2}, 25, 27\frac{1}{2}, 35;\ \{24, 13\})$.
10. Note that this notation does not represent the actual coalitions that form. For example, the first coalition structure means that $\{2, 3\}$ forms first, after being promised some compensation from Player 1.
11. The reader is referred to Maschler (1978, 1992) for further discussions on experimental findings.

REFERENCES

Kahan, J.P. and Rapoport, A. (1984), *Theories of Coalition Formation*. New Jersey: Lawrence Erlbaum Associates Hillsdale.

Maschler, M. (1963), The power of a coalition, *Management Science* 10, 8–29.

Maschler, M. (1978), Playing an n-person game, an experiment, in H. Sauermann (ed.), *Coalition Forming Behavior* (Contributions to Experimental Economics, Vol. 8), Tübingen: J.C.B. Mohr (Paul Siebeck), 231–328.

Maschler, M. (1992), The bargaining set, kernel and nucleolus, in R.J. Aumann and S. Hart (eds.), *Handbook of Game Theory*, Vol. 1. Amsterdam: Elsevier Science Publishers, 591–667.

Maschler, M. and Peleg, B. (1966), A characterization, existence proof and dimension bounds for the kernel of a game, *Pacific Journal of Mathematics* 18, 289–328.

Maschler, M., Potters, J.A.M. and Tijs, S.H. (1992), The general nucleolus and the reduced game property, *International Journal of Game Theory* 21, 85–106.

Owen, G. (1977), A generalization of Kohlberg Criterion, *International Journal of Game Theory* 6, 249–255.

Potters, J.A.M. and Tijs, S.H. (1992), The nucleolus of a matrix game and other nucleoli, *Mathematics of Operations Research*, Vol. 17, 164–174.

Rapoport, A. and Kahan, J.P. (1979), Standards of fairness in 4-person monopolistic cooperative games, in S.J. Brams, A. Schotter and G. Schwödiauer (eds), *Applied Game Theory*, Würzburg: Physica-Verlag.

Rapoport A. and Kahan, J.P. (1982), The power of a coalition and payoff disbursement in 3-person negotiable conflicts, *Journal of Mathematical Sociology* 8, 193–224.

Schmeidler, D. (1969), The nucleolus of a characteristic function game, *SIAM Journal of Applied Mathematics* 17, 1163–1170.

Shapley, L.S. (1953), Quota solutions of n-person games, in H. Kuhn and A.W. Tucker (eds), *Contribution to the Theory of Games*, Vol. 2. Princeton: Princeton University Press, 307–317.

Wallmeier, E. (1980), Der f-Nukleolus als Lösungkonsept für n-Personenspiele in Funktionform. Diplomatarbeit, Universität Münster.

Address for correspondence: Michael Maschler, Department of Mathematics and The Center for Rationality and Interactive Decision Theory, The Hebrew University of Jerusalem, 91904 Jerusalem, Israel.
E-mail: Maschler@vms.huji.ac.il

SERGIU HART

A COMPARISON OF NON-TRANSFERABLE UTILITY VALUES*

ABSTRACT. Three values for non-transferable utility games – the Harsanyi NTU-value, the Shapley NTU-value, and the Maschler–Owen consistent NTU-value – are compared in a simple example.

KEY WORDS: cooperative games, non-transferable utility (NTU), value, NTU-value, Shapley value, Harsanyi value, Maschler–Owen value, consistent value

1. INTRODUCTION

A general *non-transferable utility game in coalitional form* – an *NTU-game* for short – is given by its set of players and the sets of outcomes that are feasible for each subset ("coalition") of players.

A central solution concept for coalitional games is that of *value*, originally introduced by Shapley (1953) for games with *transferable utility* (or *TU-games* for short). For *pure bargaining problems* – where only the grand coalition of all players is essential – the classical solution is the Nash (1950) *bargaining solution*. Since the TU-games and the pure bargaining problems are two special classes of NTU-games, one looks for an NTU-solution that extends both.

An *NTU-value* is thus a solution concept for NTU-games that satisfies the following: first, it coincides with the Shapley TU-value for TU-games; second, it coincides with the Nash bargaining solution for pure bargaining problems; and third, it is covariant with individual payoff rescalings (i.e., multiplying

* Dedicated to Guillermo Owen on his sixty-fifth birthday. Research partially supported by a grant of the Israel Academy of Sciences and Humanities. The author thanks Robert J. Aumann and Andreu Mas-Colell for their comments.

the payoffs of a player by a factor $\alpha > 0$ multiplies his value payoffs by the same factor α).

The above three requirements do not however determine the NTU-value uniquely. Indeed, different NTU-values have been proposed in the literature; the most notable are due to Harsanyi (1963), Shapley (1969), and Maschler and Owen (1992).[1]

In this short note we analyze in detail a simple example of an NTU-game where the three values yield different outcomes. It is essentially the simplest possible example: there are just three players (a two-player game is a pure bargaining problem), and the coalitional function corresponds to a TU-game except for one coalition, for which the transfers of utility are possible albeit at a rate different from 1. The difference between the values will be seen to be due to the way that subcoalitions are handled.[2] It is to be hoped that the analysis here will shed further light on the NTU-values, their meanings and interpretations.

The reader is referred to the chapters on value in the *Handbook of Game Theory*, in particular McLean (2002), for further material and references.

The game is defined in Section 2; the values are computed in Sections 3–5, and compared in Section 6. Sections 7 and 8 present an exchange economy (a "market") and a "prize game" (see Hart (1994)) that generate our example.

Some notations: $\mathbb{R}$ is the real line; for a finite set S, the number of elements of S is denoted $|S|$; the $|S|$-dimensional Euclidean space with coordinates indexed by S (or, equivalently, the set of real functions on S) is $\mathbb{R}^S$; the nonnegative orthant of $\mathbb{R}^S$ is $\mathbb{R}^S_+$; and $A \subset B$ denotes *weak* inclusion (i.e., possibly $A = B$).

2. THE EXAMPLE

A *non-transferable utility game in coalitional form* is a pair (N, V), where N – the set of *players* – is a finite set, and V – the *coalitional function* – is a mapping that associates to each coalition $S \subset N$ the set $V(S) \subset \mathbb{R}^S$ of *feasible payoff vectors* for S. An element $x = (x^i)_{i \in S}$ of $V(S)$ is interpreted as follows: there exists an outcome that is feasible for the coalition S whose utility to player i is x^i (for each i in S). Thus $V(S)$ is the set of utility combinations that are feasible for the coalition S. The standard assumptions are that for each nonempty coalition S, the set $V(S)$ is a nonempty strict subset of $\mathbb{R}^S$ that is closed, convex, and comprehensive ($y \leq x \in V(S)$ implies $y \in V(S)$).

A game (N, V) is a *transferable utility game* (or *TU-game* for short) if for each coalition S there exists a real number $v(S)$ such that $V(S) = \{x \in \mathbb{R}^S : \sum_{i \in S} x^i \leq v(S)\}$. This game is denoted (N, V) or (N, v) interchangeably, and the function v is called the *worth* function.

Our example is the NTU-game (N, V) with three players $N = \{1, 2, 3\}$ and coalitional function[3]

$$\begin{aligned} V(i) &= \{x^i : x^i \leq 0\} \quad \text{for } i = 1, 2, 3, \\ V(12) &= \{(x^1, x^2) : x^1 + x^2 \leq 36,\ x^1 + 2x^2 \leq 36\}, \\ V(13) &= \{(x^1, x^3) : x^1 + x^3 \leq 0\}, \\ V(23) &= \{(x^2, x^3) : x^2 + x^3 \leq 0\}, \\ V(123) &= \{(x^1, x^2, x^3) : x^1 + x^2 + x^3 \leq 36\}. \end{aligned}$$

Except for coalition $\{1, 2\}$ – whose feasible set $V(12)$ is depicted in Figure 1 – our game (N, V) coincides with a TU-game, which we denote by (N, w) or (N, W). Thus w is the worth function

$$w(S) = \begin{cases} 36, & \text{for } S = \{1, 2\}, \{1, 2, 3\}, \\ 0, & \text{otherwise,} \end{cases} \tag{1}$$

and $W(S) = \{x \in \mathbb{R}^S : \sum_{i \in S} x^i \leq w(S)\}$ for all S.

The game (N, V) is 0*-normalized* (single players get 0) and *monotonic* (if $S \subset T$ and $x \in V(S)$ then[4] $(x, 0^{T \setminus S}) \in V(T)$). The (Pareto efficient) boundary of $V(N)$, which is denoted $\partial V(N)$, is a hyperplane with slope $\lambda = (1, 1, 1)$.

This example is not new; a similar one appears in Owen (1972) (see also Hart and Mas-Colell (1996, p. 369)).

3. THE SHAPLEY NTU-VALUE

The Shapley NTU-value for a general NTU-game (N, V) is obtained by the following procedure:

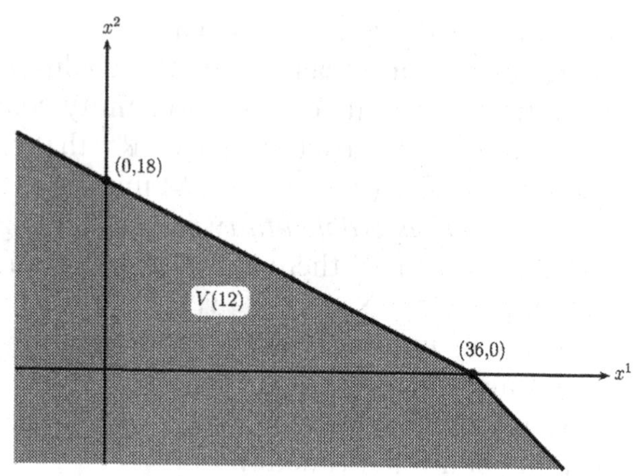

Figure 1. The feasible set for coalition {1, 2}.

- For each weight vector $\lambda \in \mathbb{R}^N_+, \lambda \neq 0$:

  1. Let the payoff vector $z \in \mathbb{R}^N$ satisfy

     $$\lambda^i z^i = \varphi^i_{\mathrm{TU}}(N, v_\lambda) \quad \text{for all } i \in N,$$

     where the TU-game (N, v_λ) is obtained from (N, V) by allowing transfers of utilities at the rates λ, i.e., $v_\lambda(S) = \sup\{\sum_{i \in S} \lambda^i x^i : (x^i)_{i \in S} \in V(S)\}$ for all[5] $S \subset N$, and φ_{TU} is the *Shapley TU-value.*
  2. If z is feasible for the grand coalition, i.e., if $z \in V(N)$, then z is a Shapley NTU-value of (N, V).

For our game (N, V) only $\lambda = (1, 1, 1)$ needs to be considered (any λ that is not a multiple of $(1, 1, 1)$ yields $v_\lambda(N) = \infty$), in which case $v_\lambda \equiv w$ (see (1)). The Shapley TU-value σ_S of each subgame[6] (S, w) is easily computed: $\sigma_{\{i\}} = 0$ for singletons, and $\sigma_{\{1,2\}} = (18, 18)$, $\sigma_{\{1,3\}} = (0, 0)$, and $\sigma_{\{2,3\}} = (0, 0)$ for the two-player subgames. It will be convenient to write σ_S as a three-dimensional vector with "–" for the players outside S:

$$\sigma_{\{1,2\}} = (18, 18, -),$$
$$\sigma_{\{1,3\}} = (0, -, 0),$$
$$\sigma_{\{2,3\}} = (-, 0, 0).$$

Next, for each two-player coalition S we adjoin to σ_S a payoff for the missing player (in boldface below) so that the resulting payoff vector $\hat{\sigma}_S$ is efficient for N (i.e., the coordinates add up to 36):

$$\hat{\sigma}_{\{1,2\}} = (18, 18, \mathbf{0}),$$
$$\hat{\sigma}_{\{1,3\}} = (0, \mathbf{36}, 0),$$
$$\hat{\sigma}_{\{2,3\}} = (\mathbf{36}, 0, 0).$$

We then average these three vectors to obtain the value for N:

$$\sigma_N = (18, 18, 0).$$

Indeed, the Shapley TU-value of a player i in a TU-game (N, v) is the average of his marginal contribution to the grand coalition $v(N) - v(N\backslash i)$, and his values in the subgames with $|N| - 1$ players:

$$\varphi^i_{\mathrm{TU}}(N, v) = \frac{1}{|N|}\left[v(N) - v(N\backslash i) + \sum_{j \in N\backslash i} \varphi^i_{\mathrm{TU}}(N\backslash j, v)\right]; \qquad (2)$$

see Hart and Mas-Colell (1996, p. 369).

The above payoff vector σ_N is thus the unique Shapley NTU-value of our game (N, V).

4. THE HARSANYI NTU-VALUE

The Harsanyi NTU-value for a general NTU-game (N, V) is obtained by the following procedure:

- For each weight vector $\lambda \in \mathbb{R}^N_+, \lambda \neq 0$:
  1. Let the payoff vector $z \in \partial V(N)$ be the *λ-egalitarian solution* of the game (N, V).
  2. If λ is a supporting normal to the boundary of $V(N)$ at z (i.e., if z is also *λ-utilitarian*) then z is a *Harsanyi NTU-value* of (N, V).

The λ-egalitarian solution is constructed recursively: for each S, given the payoff vectors $\eta_T \in V(T)$ for all strict subsets T of S, the payoff vector η_S is determined by

$$\eta_S \in \partial V(S) \quad \text{and}$$
$$\lambda^i(\eta^i_S - \eta^i_{S\backslash j}) = \lambda^j(\eta^j_S - \eta^j_{S\backslash i}) \quad \text{for all } i, j \in S. \qquad (3)$$

The λ-egalitarian solution z is the resulting payoff vector η_N for the grand coalition.

For our game (N, V) we need to consider only $\lambda = (1, 1, 1)$. This yields $\eta_{\{i\}} = 0$ for all i, and

$$\eta_{\{1,2\}} = (12, 12, -),$$

$$\eta_{\{1,3\}} = (0, -, 0),$$

$$\eta_{\{2,3\}} = (-, 0, 0).$$

Indeed, for each two-player coalition $\{i,j\}$ the egalitarian solution is the payoff vector $x \in \partial V(ij)$ with equal coordinates (i.e., $x^i = x^j$).

For the grand coalition N we use the same construction as for the Shapley TU-value. First, we extend each η_S to a payoff vector $\hat{\eta}_S$ that is efficient for N:

$$\hat{\eta}_{\{1,2\}} = (12, 12, \mathbf{12}),$$

$$\hat{\eta}_{\{1,3\}} = (0, \mathbf{36}, 0),$$

$$\hat{\eta}_{\{2,3\}} = (\mathbf{36}, 0, 0);$$

and then we average the three vectors to yield η_N:

$$\eta_N = (16, 16, 4).$$

This construction is correct since the egalitarian solution of (N, V) is the Shapley TU-value of the game (N, u) with $u(S) = \Sigma_{i \in S} \eta^i_S$ for all $S \subset N$; see Hart (1985, (4.6)).

Thus η_N is the egalitarian solution for N, and therefore the unique Harsanyi NTU-value of (N, V); in terms of Hart (1985), the collection $(\eta_S)_{S \subset N}$ we have obtained is the Harsanyi *payoff configuration.*

5. THE MASCHLER–OWEN CONSISTENT NTU-VALUE

The Maschler–Owen consistent NTU-value for a general NTU-game (N, V) is obtained recursively by the following procedure:

- Let S be a coalition, and assume that a payoff vector $\gamma_T \in V(T)$ is given for all strict subcoalitions T of S. For each weight vector $\lambda \in \mathbb{R}^S_+, \lambda \neq 0$:

1. Let the payoff vector $z \in \mathbb{R}^S$ satisfy

$$\lambda^i z^i = \frac{1}{|S|}\left[v_\lambda(S) - \sum_{j\in S\setminus i} \lambda^j \gamma^j_{S\setminus i} + \sum_{j\in S\setminus i} \lambda^i \gamma^i_{S\setminus j}\right] \quad \text{for all } i \in S, \tag{4}$$

where, again, $v_\lambda(S) = \sup\{\sum_{i\in S} \lambda^i x^i : (x^i)_{i\in S} \in V(S)\}$.
2. If z is feasible for the coalition S, i.e., if $z \in V(S)$, then define $\gamma_S = z$.

- The resulting payoff vector γ_N for the grand coalition is then a *Maschler–Owen consistent NTU-value* of (N, V).

Formula (4) – which is a generalization of (2) in the TU-case – is equivalent to[7]

$$\sum_{i\in S} \lambda^i z^i = v_\lambda(S) \quad \text{and}$$

$$\sum_{j\in S\setminus i} \lambda^i (z^i - \gamma^i_{S\setminus j}) = \sum_{j\in S\setminus i} \lambda^j (z^j - \gamma^j_{S\setminus i}) \quad \text{for all } i \in S;$$

see Proposition 4 and Formula (3) in Hart and Mas-Colell (1996), and Formula (5.1) in Maschler and Owen (1989).

For our game (N, V) we have $\gamma_{\{i\}} = 0$ for all i, and

$$\gamma_{\{1,2\}} = (18, 9, -),$$
$$\gamma_{\{1,3\}} = (0, -, 0),$$
$$\gamma_{\{2,3\}} = (-, 0, 0).$$

Indeed, for two players, the consistent value coincides with the Nash bargaining solution.

Now (4) allows us to use again the "extension" construction ($\lambda = (1, 1, 1)$):

$$\hat{\gamma}_{\{1,2\}} = (18, 9, \mathbf{9}),$$
$$\hat{\gamma}_{\{1,3\}} = (0, \mathbf{36}, 0),$$
$$\hat{\gamma}_{\{2,3\}} = (\mathbf{36}, 0, 0),$$

and their average is

$$\gamma_N = (18, 15, 3).$$

Therefore γ_N is the unique Maschler–Owen consistent NTU-value of (N, V).

6. A COMPARISON

To recapitulate, the three NTU-values of our game (N, V) are

$$\begin{aligned} \varphi^{\text{Sh}}(N, V) &= (18, 18, 0), \\ \varphi^{\text{Ha}}(N, V) &= (16, 16, 4), \\ \varphi^{\text{MO}}(N, V) &= (18, 15, 3). \end{aligned} \tag{5}$$

The computations leading to these values clearly exhibit that the difference between them derives from the way the intermediate payoff vector $x_{\{1,2\}}$ for the coalition $\{1, 2\}$ – the only coalition whose feasible set is *not* of the TU-type – is determined. Indeed, the payoff vectors x_S for all other strict subsets of N are identical for the three values, and, once all the x_S are given, the value for the grand coalition x_N is uniquely determined (by the "extension" construction with respect to $\lambda = (1, 1, 1)$, the unique supporting normal to $\partial V(N)$).

The payoff vectors $x_{\{1,2\}}$ for the coalition $\{1, 2\}$ are, respectively,

$$\begin{aligned} x^{\text{Sh}}_{\{1,2\}} &= (18, 18), \\ x^{\text{Ha}}_{\{1,2\}} &= (12, 12), \\ x^{\text{MO}}_{\{1,2\}} &= (18, 9); \end{aligned}$$

(see Figure 2). The Shapley NTU-value and the Harsanyi NTU-value both take $x_{\{1,2\}}$ to be an egalitarian outcome (i.e., an "equal-split" payoff vector – since the rates of interpersonal utility comparison λ dictated by the grand coalition satisfy $\lambda^1 = \lambda^2$). The difference is that the Harsanyi approach uses $V(12)$, the feasible set for $\{1, 2\}$, to determine $x_{\{1,2\}}$, whereas the Shapley approach allows transfers of utility at the rates λ of the grand coalition and so $V(12)$ is replaced by $W(12)$ (which corresponds to $v_\lambda(12)$). Thus $x^{\text{Ha}}_{\{1,2\}} = (12, 12) \in \partial V(12)$ and $x^{\text{Sh}}_{\{1,2\}} = (18, 18) \in \partial W(12)$. As for the Maschler–Owen NTU-value, it considers coalition $\{1, 2\}$ independently of the grand

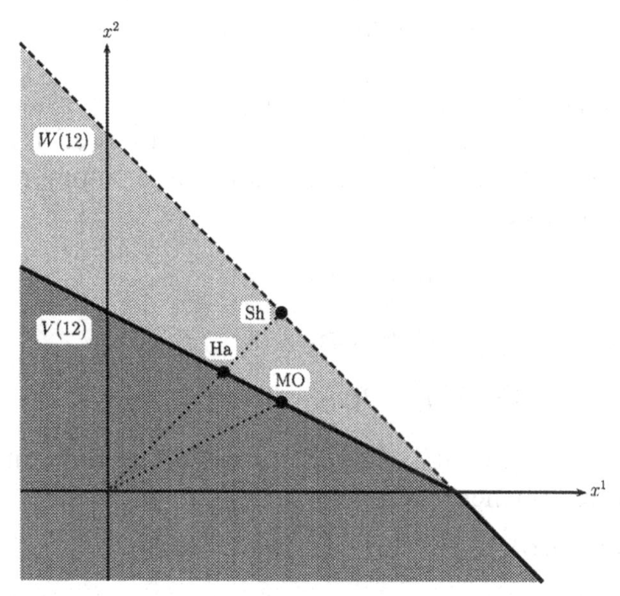

Figure 2. The intermediate payoff vectors for coalition {1, 2}.

coalition: $x_{\{1,2\}}$ is determined by the $\{1,2\}$-subgame only. Moreover, $x_{\{1,2\}}$ is determined for $\{1,2\}$ in exactly the same way that x_N is determined for N; this property – that x_S is the consistent NTU-value of the S-subgame for each S – is called "subcoalition perfectness" in Hart and Mas-Colell (1996, p. 366). Thus $x^{\mathrm{MO}}_{\{1,2\}} = (18, 9)$, the Nash bargaining solution of the two-person game.

Which approach is "correct"? There cannot be a definite answer.[8] For instance, it may depend on the way the interactions between the players are conducted.[9] If transfers are allowed (or "implied" by the grand coalition[10]) as in the Shapley NTU-value, then player 3 becomes a null ("dummy") player, and his value of 0 is justified. Otherwise player 3 is not a null player, and his value is positive. The Harsanyi NTU-value is egalitarian-based; therefore players 1 and 2 get equal payoffs. In contrast, the Maschler–Owen NTU-value takes into account the asymmetry between the two players in the subcoalition $\{1,2\}$ – and is the only one to do so. Thus it appears that the Maschler–Owen consistent NTU-value reflects the structure of this game better than the other NTU-values.[11]

7. AN EXCHANGE ECONOMY

Our example is essentially a market game.[12] For instance, take E to be the following exchange economy ("market"): there are three players ("traders") $i = 1, 2, 3$, and three commodities; the utility functions are

$$u^1(a_1, a_2, a_3) = 36a_1 + 36a_2 - 36,$$

$$u^2(a_1, a_2, a_3) = 18a_1 + 36a_3,$$

$$u^3(a_1, a_2, a_3) = 36a_1 + 36a_3 - 36$$

(a_j denotes the quantity of good j), and the initial commodity bundles ("endowments") are

$$e^1 = (1, 0, 0),$$

$$e^2 = (0, 1, 0),$$

$$e^3 = (0, 0, 1).$$

Let (N, V^E) be the resulting *NTU-market game*; i.e., $V^E(S) = \{x \in \mathbb{R}^S$: there exists an S-allocation $(c^i)_{i \in S}$ with $\Sigma_{i \in S}\, c^i = \Sigma_{i \in S}\, e^i, c^i \in \mathbb{R}^3_+$ and $x^i \leq u^i(c^i)$ for all $i \in S\}$ for all $S \subset N = \{1, 2, 3\}$. The individually rational payoff vectors of (N, V^E) coincide with those of our example (N, V); i.e., for all $S \subset N$ we have $V^E(S) \cap \mathbb{R}^S_+ = V(S) \cap \mathbb{R}^S_+$ (note that $u^i(e^i) = 0$ and $\partial V(i) = \{0\}$ for all i), and also $V^E(S) \subset V(S)$. This implies that the NTU-values of (N, V) given in (5) are also NTU-values for (N, V^E). One can check that (N, V^E) has no other values.[13]

8. A PRIZE GAME

Our example is also essentially a *hyperplane game*, and it can thus be represented as a *prize game*; see Hart (1994). Indeed, let the prize of the grand coalition be worth 36 to each player, and let the prize of coalition $\{1, 2\}$ be worth 36 to player 1 and 18 to player 2 (there are no other prizes). The resulting game (N, V^*) is again identical to our example (N, V) in the individually rational region, and its NTU-values are given by (5).

NOTES

1. The Shapley NTU-value is sometimes referred to as the 'λ-transfer value,' and the Maschler–Owen value is called the 'consistent NTU-value.' Axiomatizations of these values have been provided by Aumann (1985) for the Shapley NTU-value, by Hart (1985) for the Harsanyi NTU-value, by de Clippel, Peters and Zank (2002) and Hart (1994, 2003) for the Maschler–Owen NTU-value. Another NTU-value was proposed by Owen (1972).
2. See also the discussions in Hart (1985, Section 5), Hart and Mas-Colell (1996, Section 4), and de Clippel et al. (2002, Section 4).
3. For simplicity we write $V(1)$, $V(12), \ldots$ instead of the more cumbersome $V(\{1\})$, $V(\{1,2\}), \ldots$; similarly, $S\backslash i$ for $S\backslash\{i\}$, and so on.
4. $0^{T\backslash S}$ is the 0-vector in $\mathbb{R}^{T\backslash S}$.
5. If the 'sup' in the definition $v_\lambda(S)$ is infinite for some S then there is no NTU-value corresponding to this λ (and the procedure for this λ stops here).
6. The 'subgame' (S, V) of (N, V) is obtained by restricting the domain of V to the subsets of S.
7. Compare (3).
8. For a general discussion of the multiplicity of solution concepts, see '1930–1950, Section iii' in Aumann (1987).
9. Different NTU-values may be thought of, inter alia, as corresponding to different bargaining procedures – from which the coalitional form abstracts away. For example, the noncooperative model of Hart and Mas-Colell (1996) leads to the Maschler–Owen NTU-value (and thus, a fortiori, to the Shapley TU-value and the Nash bargaining solution). It would be of interest to obtain explicit bargaining procedures leading to other NTU-values.
10. See Myerson (1991, pp. 475–476).
11. To emphasize the differences between the values, consider the case where $V(12)$ becomes more and more "flat": replace $x^1 + 2x^2 \leq 36$ with $x^1 + mx^2 \leq 36$, and let $m \to \infty$. Then $\varphi^{\mathrm{Sh}} = (18, 18, 0)$, $\varphi^{\mathrm{Ha}} \to (12, 12, 12)$, and $\varphi^{\mathrm{MO}} \to (18, 12, 6)$.
12. I.e., it coincides with a market game in the relevant (individually rational) region; see below.
13. Since $\lambda = (1, 1, 1)$ is no longer the unique supporting normal to the boundary of $V^E(N)$, one needs to consider additional weight vectors λ (including zero weights). We omit the straightforward but lengthy arguments that show that no other values are obtained.

REFERENCES

Aumann, R.J. (1985), An axiomatization of the non-transferable utility value, *Econometrica* 53, 599–612.

Aumann, R.J. (1987), Game theory, in J. Eatwell, M. Milgate, and P. Newman (eds), *The New Palgrave: A Dictionary of Economics*, Vol. 2, London: Macmillan, 460–482.

De Clippel, G., Peters, H. and Zank, H. (2002), Axiomatizing the Harsanyi Solution, the Symmetric Egalitarian Solution, and the Consistent Shapley Solution for NTU-Games, mimeo.

Harsanyi, J.C. (1963), A simplified bargaining model for the *n*-person cooperative game, *International Economic Review* 4, 194–220.

Hart, S. (1985), An axiomatization of Harsanyi's non-transferable utility solution, *Econometrica* 53, 1295–1313.

Hart, S. (1994), On prize games, in N. Megiddo (ed.), *Essays in Game Theory*, Berlin: Springer-Verlag, 111–121.

Hart, S. (2003), An Axiomatization of the Consistent Non-Transferable Utility Value, Center for Rationality DP-337, The Hebrew University of Jerusalem, mimeo.

Hart, S. and Mas-Colell, A. (1996), Bargaining and value, *Econometrica* 64, 357–380.

Maschler, M. and Owen, G. (1989), The consistent Shapley value for hyperplane games, *International Journal of Game Theory* 18, 389–407.

Maschler, M. and Owen, G. (1992), The consistent Shapley value for games without side payments, in R. Selten (ed.), *Rational Interaction*, New York: Springer-Verlag, 5–12.

McLean, R. (2002), Values of non-transferable utility games, in R.J. Aumann and S. Hart (eds), *Handbook of Game Theory, with Economic Applications*, Vol. 3, Amsterdam: Elsevier, 2077–2120.

Myerson, R.B. (1991), *Game Theory*, Cambridge MA: Harvard University Press.

Nash, J. (1950), The bargaining problem, *Econometrica* 18, 155–162.

Owen, G. (1972), A value for games without side payments, *International Journal of Game Theory* 1, 95–109.

Shapley, L.S. (1953), A value for *n*-person games, in H.W. Kuhn and A.W. Tucker (eds), *Contributions to the Theory of Games II, Annals of Mathematics Studies 28*, Princeton: Princeton University Press, 307–317.

Shapley, L.S. (1969), Utility comparison and the theory of games, in *La Décision*, Paris: Editions du CNRS, 251–263.

Address for correspondence: Sergiu Hart, Center for Rationality and Interactive Decision Theory, Department of Mathematics, and Department of Economics, The Hebrew University of Jerusalem, Feldman Building, Givat Ram, 91904 Jerusalem, Israel (E-mail: hart@huji.ac.il;
URL: http://www.ma.huji.ac.il/~hart)

RODICA BRANZEI, STEFANO MORETTI, HENK NORDE
and STEF TIJS

THE *P*-VALUE FOR COST SHARING IN MINIMUM COST SPANNING TREE SITUATIONS

ABSTRACT. The aim of this paper is to introduce and axiomatically characterize the *P*-value as a rule to solve the cost sharing problem in minimum cost spanning tree (mcst) situations. The *P*-value is related to the Kruskal algorithm for finding an mcst. Moreover, the *P*-value leads to a core allocation of the corresponding mcst game, and when applied also to the mcst subsituations it delivers a population monotonic allocation scheme. A cone-wise positive linearity property is one of the basic ingredients of an axiomatic characterization of the *P*-value.

KEY WORDS: Cost sharing, Minimum cost spanning tree games, Population monotonic allocation schemes, Value

1. INTRODUCTION

The topic in this paper lies on the borderline of Operations Research and Game Theory, an inspiring place also for Guillermo Owen (cf. Owen, 1975, 1995, 1999; Gambarelli and Owen, 1994). Since the basic paper of Bird (1976) much attention has been paid to the problem of sharing costs in situations where agents have to be connected with a source as cheap as possible, and where connections between users and between users and the source can be shared among users if they cooperate. Let us refer to the dissertations of Aarts (1994) and Feltkamp (1995), and to the papers of Granot and Huberman (1981), Feltkamp et al. (1994), and Kar (2002). In the papers of Dutta and Kar (2002), Kent and Skorin-Kapov (1996), Moretti et al. (2002), and Norde et al. (2004), the existence of cost monotonic and population monotonic allocation rules (Sprumont, 1990) is central.

The *P*-value introduced in Section 3, has been arisen from our interest in monotonic allocation schemes too (Tijs et al.,

Theory and Decision **56:** 47–61, 2004.

2003). Our introduction of the P-value is a two-step procedure. First, we define this value on cones of mcst situations with the same ordering pattern of the edges with respect to costs. Then, we prove that we can patch these P-values together to the whole cone of mcst situations. It turns out that our P-value equals the Equal Remaining Obligations (ERO) rule suggested by Jos Potters (which explains the name of our rule) and which is studied first in Feltkamp et al. (1994). Furthermore, our P-value turns out to be the average of the population monotonic allocation rules introduced in Norde et al. (2004). In Section 4 we give an axiomatic characterization of the P-value, where the cone-wise positive linearity of P is a fundamental property and where the decomposition of an mcst situation into simple mcst situations (cf. Kuipers, 1993; Norde et al., 2004) plays a role. In Section 5, which concludes the paper, the related Π-value for minimum spanning tree games is introduced and it is shown that the Π-value is a population monotonic allocation rule.

2. PRELIMINARIES AND NOTATIONS

First, we recall some definitions from graph theory which are used in this paper. An (undirected) *graph* is a pair $\langle V, E \rangle$, where V is a set of vertices or nodes and E is a set of edges e of the form $\{i,j\}$ with $i,j \in V$, $i \neq j$. The *complete graph* on a set V of vertices is the graph $\langle V, E_V \rangle$, where $E_V = \{\{i,j\} | i,j \in V, i \neq j\}$. A *path* between i and j in a graph $\langle V, E \rangle$ is a sequence of nodes $i = i_0, i_1, \ldots, i_k = j$, $k \geq 1$, such that $\{i_s, i_{s+1}\} \in E$ for each $s \in \{0, \ldots, k-1\}$. A *cycle* in $\langle V, E \rangle$ is a path from i to i for some $i \in V$. Two nodes $i,j \in V$ are connected in (V, E) if $i = j$ or if there exists a path between i and j in E.

Now, we consider *minimum cost spanning tree* (mcst) *situations*. In an mcst situation a set $N = \{1, \ldots, n\}$ of agents is involved willing to be connected as cheap as possible to a source (i.e. a supplier of a service) denoted by 0. In the sequel we use the notation $N' = N \cup \{0\}$. An mcst situation can be represented by a tuple $\langle N', E_{N'}, w \rangle$, where $\langle N', E_{N'} \rangle$ is the complete graph on the set N' of nodes or vertices, and $w : E_{N'} \to \mathbb{R}_+$ is a map which assigns to each edge $e \in E_{N'}$ a non-negative number $w(e)$ representing the weight or cost of edge e. We call w a *weight function*. If $w(e) \in \{0, 1\}$ for every $e \in E_{N'}$, the weight function w is called a *simple weight function*, and we refer then to $\langle N', E_{N'}, w\rangle$ as a *simple mcst situation*.

Since in our paper the graph of possible edges is always the complete graph, we simply denote an mcst situation with set of users N, source 0, and weight function w by $\langle N', w\rangle$. Often we identify an mcst situation $\langle N', w\rangle$ with the corresponding weight function w. We denote by $\mathcal{W}^{N'}$ the set of all mcst situations $\langle N', w\rangle$ (or w) with node set N'. For each $S \subseteq N$, one can consider the mcst subsituation $\langle S', w_{|S'}\rangle$, where $S' = S \cup \{0\}$ and $w_{|S'} : E_{S'} \to \mathbb{R}_+$ is the restriction of the weight function w to $E_{S'} \subseteq E_{N'}$, i.e. $w_{|S'}(e) = w(e)$ for each $e \in E_{S'}$.

Let $\langle N', w\rangle$ be an mcst situation. Two nodes i and j are called (w, N')-connected if $i = j$ or if there exists a sequence of nodes $i = i_0, \ldots, i_k = j$ in N', $k \geq 1$, with $w(\{i_s, i_{s+1}\}) = 0$ for every $s \in \{0, \ldots, k-1\}$. A (w, N')*-component* of N' is a maximal subset of N' with the property that any two nodes in this subset are (w, N')-connected. We denote by $C_i(w)$ the (w, N')-component to which i belongs and by $\mathcal{C}(w)$ the set of all (w, N')-components of N'. Clearly, the collection of (w, N')-components forms a partition of N'.

We define the set $\Sigma_{E_{N'}}$ of *linear orders* on $E_{N'}$ as the set of all bijections $\sigma : \{1, \ldots, |E_{N'}|\} \to E_{N'}$, where $|E_{N'}|$ is the cardinality of the set $E_{N'}$. For each mcst situation $\langle N', w\rangle$ there exists at least one linear order $\sigma \in \Sigma_{E_{N'}}$ such that $w(\sigma(1)) \leq w(\sigma(2)) \leq \cdots \leq w(\sigma(|E_{N'}|))$. We denote by w^σ the column vector $(w(\sigma(1)), w(\sigma(2)), \ldots, w(\sigma(|E_{N'}|)))^t$.

For any $\sigma \in \Sigma_{E_{N'}}$ we define the set

$$K^\sigma = \{w \in \mathbb{R}_+^{E_{N'}} | w(\sigma(1)) \leq w(\sigma(2)) \leq \cdots \leq w(\sigma(|E_{N'}|))\}.$$

The set K^σ is a cone in $\mathbb{R}_+^{E_{N'}}$, which we call the *Kruskal cone with respect to* σ. One can easily see that $\cup_{\sigma \in \Sigma_{E_{N'}}} K^\sigma = \mathbb{R}_+^{E_{N'}}$. For each $\sigma \in \Sigma_{E_{N'}}$ the cone K^σ is a simplicial cone with generators $e^{\sigma,k} \in K^\sigma$, $k \in \{1, 2, \ldots, |E_{N'}|\}$, where

$$e^{\sigma,k}(\sigma(1)) = e^{\sigma,k}(\sigma(2)) = \cdots = e^{\sigma,k}(\sigma(k-1)) = 0 \quad \text{and}$$
$$e^{\sigma,k}(\sigma(k)) = e^{\sigma,k}(\sigma(k+1)) = \cdots = e^{\sigma,k}(\sigma(|E_{N'}|)) = 1. \tag{1}$$

[Note that $e^{\sigma,1}(\sigma(k)) = 1$ for all $k \in \{1, 2, \ldots, |E_{N'}|\}$.]

This implies that each $w \in K^\sigma$ can be written in a unique way as non-negative linear combination of these generators. To be more concrete, for $w \in K^\sigma$ we have

$$w = w(\sigma(1))e^{\sigma,1} + \sum_{k=2}^{|E_{N'}|} (w(\sigma(k)) - w(\sigma(k-1)))e^{\sigma,k}. \tag{2}$$

Clearly, we can also write $\mathcal{W}^{N'} = \cup_{\sigma \in \Sigma_{E_{N'}}} K^\sigma$, if we identify an mcst situation $\langle N', w \rangle$ with w.

Any mcst situation gives rise to two problems: the construction of a network $\Gamma \subseteq E_{N'}$ of minimal cost connecting all users to the source, and a cost sharing problem of distributing this cost among users in a fair way. The cost of a network Γ is $w(\Gamma) = \sum_{e \in \Gamma} w(e)$. A network Γ is a *spanning network* on $S' \subseteq N'$ if for every $e \in \Gamma$ we have $e \in E_{S'}$ and for every $i \in S$ there is a path in Γ from i to the source. The cost of a minimum (cost) spanning network Γ on N' in a simple mcst situation equals $|C(w)| - 1$ (see Lemma 2 in Norde et al., 2004). To construct a minimum cost spanning network Γ on N' we use in this paper the Kruskal algorithm (Kruskal, 1956), where the edges are considered one by one according to non-decreasing cost, and an edge is either rejected, if it generates a cycle with the edges already constructed, or it is constructed, otherwise.

Let $\langle N', w \rangle$ be an mcst situation. The *minimum cost spanning tree game* (N, c_w) (or simply c_w), corresponding to $\langle N', w \rangle$, is defined by

$$c_w(S) = \min\{w(\Gamma) | \Gamma \text{ is a spanning network on } S'\},$$

for every $S \in 2^N \backslash \{\emptyset\}$, where 2^N stands for the power set of the player set N, with the convention that $c_w(\emptyset) = 0$.

We denote by $\mathcal{MCST}^N$ the class of all mcst games corresponding to mcst situations in $\mathcal{W}^{N'}$. For each $\sigma \in \Sigma_{E_{N'}}$, we denote by $\mathcal{G}^\sigma$ the set $\{c_w | w \in K^\sigma\}$ which is a cone. We can express

$\mathcal{MCST}^N$ as the union of all cones $\mathcal{G}^\sigma$, i.e. $\mathcal{MCST}^N = \cup_{\sigma \in \Sigma_{E_{N'}}} \mathcal{G}^\sigma$, and we would like to point out that $\mathcal{MCST}^N$ itself is not a cone if $|N| \geq 2$.

3. THE P-VALUE

Let $w \in \mathcal{W}^{N'}$ and let $\sigma \in \Sigma_{E_{N'}}$ be such that $w \in K^\sigma$. We can consider a sequence of precisely $|E_{N'}| + 1$ graphs $\langle N', F^{\sigma,0} \rangle$, $\langle N', F^{\sigma,1} \rangle, \ldots, \langle N', F^{\sigma,|E_{N'}|} \rangle$ such that $F^{\sigma,0} = \emptyset$, $F^{\sigma,k} = F^{\sigma,k-1} \cup \{\sigma(k)\}$ for each $k \in \{1, \ldots, |E_{N'}|\}$. Now, we define the *connection vectors* $b^{\sigma,k} \in \mathbb{R}^N$ for each $k \in \{0, 1, \ldots, |E_{N'}|\}$ as follows:

$$b_i^{\sigma,k} = \begin{cases} 0 & \text{if } i \text{ is connected to 0 in } \langle N', F^{\sigma,k} \rangle, \\ \frac{1}{n_i(F^{\sigma,k})} & \text{otherwise,} \end{cases} \tag{3}$$

for each $i \in N$, where $n_i(F^{\sigma,k})$ is the number of nodes in N which are connected to i, directly or indirectly, via edges in $F^{\sigma,k}$. Note that $n_i(F^{\sigma,k}) = 1$ if i is disconnected from each other node in $\langle N', F^{\sigma,k} \rangle$. Note also that for each $\sigma \in \Sigma_{E_{N'}}$, $b_i^{\sigma,0} = 1$ and $b_i^{\sigma,|E_{N'}|} = 0$, for each $i \in N$.

EXAMPLE 1. Consider the mcst situation $\langle N', w \rangle$ with $N' = \{0, 1, 2, 3\}$ and w as depicted in Figure 1. Note that $w \in K^\sigma$, with $\sigma(1) = \{1, 3\}$, $\sigma(2) = \{1, 2\}$, $\sigma(3) = \{2, 3\}$, $\sigma(4) = \{1, 0\}$, $\sigma(5) = \{2, 0\}$, $\sigma(6) = \{3, 0\}$.

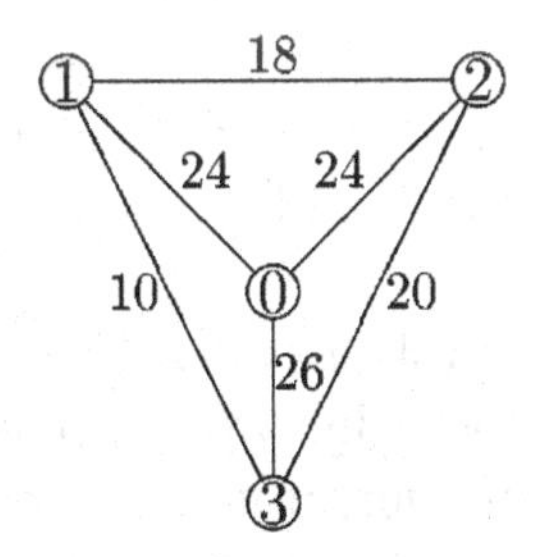

Figure 1. An mcst situation with three agents.

The sequence of seven graphs $\langle N', F^{\sigma,k} \rangle$ and the corresponding connection vectors $b^{\sigma,k}$ are shown in the following Table:

$\langle N',\emptyset\rangle$	$b^{\sigma,0} = (1,1,1)^t$
$\langle N',\{\{1,3\}\}\rangle$	$b^{\sigma,1} = \left(\frac{1}{2},1,\frac{1}{2}\right)^t$
$\langle N',\{\{1,3\},\{1,2\}\}\rangle$	$b^{\sigma,2} = \left(\frac{1}{3},\frac{1}{3},\frac{1}{3}\right)^t$
$\langle N',\{\{1,3\},\{1,2\},\{2,3\}\}\rangle$	$b^{\sigma,3} = \left(\frac{1}{3},\frac{1}{3},\frac{1}{3}\right)^t$
$\langle N',\{\{1,3\},\{1,2\},\{2,3\},\{1,0\}\}\rangle$	$b^{\sigma,4} = (0,0,0)^t$
$\langle N',\{\{1,3\},\{1,2\},\{2,3\},\{1,0\},\{2,0\}\}\rangle$	$b^{\sigma,5} = (0,0,0)^t$
$\langle N',\{\{1,3\},\{1,2\},\{2,3\},\{1,0\},\{2,0\},\{3,0\}\}\rangle$	$b^{\sigma,6} = (0,0,0)^t$

Remark 1. Let $\sigma \in \Sigma_{E_{N'}}$. For each $k \in \{1,\ldots,|E_{N'}|\}$, consider the simple mcst situation $e^{\sigma,k}$. Then, for $k > 1$, each edge $e \in F^{\sigma,k-1}$ has cost $e^{\sigma,k}(e) = 0$. Therefore, if i and j in N' are connected in $\langle N', F^{\sigma,k-1}\rangle$, then they are also in the same $(e^{\sigma,k}, N')$-component. Conversely, if i and j are in the same $(e^{\sigma,k}, N')$-component, then they are also connected in $\langle N', F^{\sigma,k-1}\rangle$ and as a consequence, by relation (3), $b_i^{\sigma,k-1} = b_j^{\sigma,k-1}$.

DEFINITION 1. Let $\sigma \in \Sigma_{E_{N'}}$. The contribution matrix w.r.t. σ is the matrix $M^\sigma \in \mathbb{R}^{N\times E_{N'}}$ where the rows correspond to the agents and the columns to the edges, and such that the kth column of M^σ equals

$$M^\sigma e^k = b^{\sigma,k-1} - b^{\sigma,k}, \tag{4}$$

for each $k \in \{1,\ldots,|E_{N'}|\}$. [Here e^k is a column vector such that $e_i^k = 1$ if $i = k$ and $e_i^k = 0$ for each $i \in \{1,\ldots,|E_{N'}|\}\setminus\{k\}$.]

Note that each column $M^\sigma e^k$ such that $(M^\sigma e^k)_i \neq 0$ for some $i \in N$ corresponds to the edge $\sigma(k)$ constructed at stage k in Kruskal's algorithm. Further, note that the sum of the elements of such a column equals 1. The interpretation of the contribution vector $M^\sigma e^k$ is that each entry $(M^\sigma e^k)_i$, $i \in N$, represents the fraction of the cost of edge $\sigma(k)$ paid by user i. On the other hand, the zero columns in M^σ correspond to rejected edges in Kruskal's algorithm.

Another characteristic of the contribution matrix is that the sum of the elements in each row $i \in N$ is equal to 1,

$$\sum_{k=1}^{|E_{N'}|} (M^{\sigma} e^k)_i = \sum_{k=1}^{|E_{N'}|} (b_i^{\sigma,k-1} - b_i^{\sigma,k}) \\ = b_i^{\sigma,0} - b_i^{\sigma,|E_{N'}|} = 1 - 0 = 1. \tag{5}$$

DEFINITION 2. For each $\sigma \in \Sigma_{E_{N'}}$, we define the P^{σ}-value as the map $P^{\sigma}: K^{\sigma} \to \mathbb{R}^N$, where $P^{\sigma}(w) = M^{\sigma} w^{\sigma}$ for each mcst situation w in the cone K^{σ}.

In order to define the P-value on $\mathcal{W}^{N'}$ we need Proposition 1, which follows directly from the following lemma.

LEMMA 1. *Let* $\sigma \in \Sigma_{E_{N'}}, w \in K^{\sigma}$ *and suppose that, for some* $t \in \{1, \ldots, |E_{N'}| - 1\}$, $w_t^{\sigma} = w_{t+1}^{\sigma}$. *Then for the ordering* $\sigma' \in \Sigma_{E_{N'}}$ *such that* $\sigma'(\mathrm{i}) = \sigma(\mathrm{i})$ *for each* $i \in \{1, \ldots, |E_{N'}|\} \backslash \{t, t+1\}$, $\sigma'(\mathrm{t}) = \sigma(\mathrm{t}+1)$ *and* $\sigma'(\mathrm{t}+1) = \sigma(\mathrm{t})$, *we have that* $w \in K^{\sigma'}$ *and* $P^{\sigma}(w) = P^{\sigma'}(w)$.

Proof. It is obvious that $w \in K^{\sigma'}$. Put $a = w_t^{\sigma}$. Note that $b^{\sigma,k} = b^{\sigma',k}$ for all $k \in \{1, \ldots, |E_{N'}|\}$ with $k \neq t$. This implies that $w_k^{\sigma} M^{\sigma} e^k = w_k^{\sigma'} M^{\sigma'} e^k$ for all $k \in \{1, \ldots, |E_{N'}|\}$ with $k \notin \{t, t+1\}$ and

$$\begin{aligned} & w_t^{\sigma'} M^{\sigma'} e^t + w_{t+1}^{\sigma'} M^{\sigma'} e^{t+1} \\ & = a(b^{\sigma',t-1} - b^{\sigma',t}) + a(b^{\sigma',t} - b^{\sigma',t+1}) \\ & = a(b^{\sigma',t-1} - b^{\sigma',t+1}) = a(b^{\sigma,t-1} - b^{\sigma,t+1}) \\ & = a(b^{\sigma,t-1} - b^{\sigma,t}) + a(b^{\sigma,t} - b^{\sigma,t+1}) \\ & = w_t^{\sigma} M^{\sigma} e^t + w_{t+1}^{\sigma} M^{\sigma} e^{t+1}. \end{aligned} \tag{6}$$

So, $M^{\sigma} w^{\sigma} = M^{\sigma'} w^{\sigma'}$ or, equivalently, $P^{\sigma}(w) = P^{\sigma'}(w)$. □

PROPOSITION 1. *If* $w \in K^{\sigma} \cap K^{\sigma'}$ *with* σ, $\sigma' \in \Sigma_{E_{N'}}$, *then* $P^{\sigma}(w) = P^{\sigma'}(w)$.

This proposition makes it possible to define the *P-value on* $\mathcal{W}^{N'}$.

DEFINITION 3. The P-value is the map $P: \mathcal{W}^{N'} \to \mathbb{R}^N$, defined by

$$P(w) = P^{\sigma}(w) = M^{\sigma} w^{\sigma} \tag{7}$$

for each $w \in \mathcal{W}^{N'}$ and $\sigma \in \Sigma_{E_{N'}}$ such that $w \in K^{\sigma}$.

EXAMPLE 2. Consider again the mcst situation in Example 1. Then the contribution matrix is

$$M^{\sigma} = \begin{pmatrix} \frac{1}{2} & \frac{1}{6} & 0 & \frac{1}{3} & 0 & 0 \\ 0 & \frac{2}{3} & 0 & \frac{1}{3} & 0 & 0 \\ \frac{1}{2} & \frac{1}{6} & 0 & \frac{1}{3} & 0 & 0 \end{pmatrix}$$

and $w^{\sigma} = (10, 18, 20, 24, 24, 26)^t$. Therefore $P(w) = M^{\sigma} w^{\sigma} = (16, 20, 16)^t$.

An alternative way of calculating $P(w)$, which will be useful in the following, is as non-negative linear combination of $P(e^{\sigma,k})$, $k \in \{1, \ldots, |E_{N'}|\}$, where $\sigma \in \Sigma_{E_{N'}}$ is such that $w \in K^{\sigma}$ (see Equation (2)). In formula

$$P(w) = w(\sigma(1))P(e^{\sigma,1}) + \\ + \sum_{k=2}^{|E_{N'}|} (w(\sigma(k)) - w(\sigma(k-1)))P(e^{\sigma,k}). \tag{8}$$

Note that since for each $\sigma \in \Sigma_{E_{N'}}$ the connection vector $b^{\sigma,|E_{N'}|}$ is the zero vector, the P-value of each mcst situation $e^{\sigma,k} \in K^{\sigma}$, $k \in \{1, \ldots, |E_{N'}|\}$, equals the connection vector corresponding to the graph $\langle N', F^{\sigma,k-1} \rangle$

$$P(e^{\sigma,k}) = M^{\sigma} e^{\sigma,k} = \sum_{r=k}^{|E_{N'}|} (b^{\sigma,r-1} - b^{\sigma,r}) = b^{\sigma,k-1}. \tag{9}$$

Remark 2. It turns out that the P-value coincides with the *Equal Remaining Obligations* (ERO) rule. The ERO-rule has been introduced in Feltkamp et al. (1994) via an extension of Kruskal's algorithm. According to the ERO-rule at each stage $k \in \{0, 1, \ldots, |E_{N'}|\}$ of the algorithm each player $i \in N$ pays exactly the difference f_i^k between *remaining obligations* o_i^{k-1} and o_i^k, i.e. $f_i^k = o_i^{k-1} - o_i^k$ for each $i \in N$, where, as shown in Theorem 4.3 of Feltkamp et al. (1994), o_i^k is equal to $b_i^{\sigma,k}$, with σ such that $w \in K^{\sigma}$, as calculated in relation (3). An axiomatic characterization of the ERO-rule using the properties of NE (*non-emptiness*), FSC (*free-for-source-component*), LOC (*local*), Eff (*efficiency*), ET (*equal treatment*) and IPCons (*inversely proportional consistency*) is given there. In the next section we provide an alternative axiomatic characterization.

4. AN AXIOMATIC CHARACTERIZATION OF THE P-VALUE

We call a map $F : \mathcal{W}^{N'} \to \mathbb{R}^N$ assigning to every mcst situation w a unique cost allocation in $\mathbb{R}^N$ a *solution*. Some interesting properties for solutions of mcst situations are the following.

Property 1. The solution F is efficient (EFF) if for each $w \in \mathcal{W}^{N'}$

$$\sum_{i \in N} F_i(w) = w(\Gamma),$$

where Γ is a minimum cost spanning network on N'.

Property 2. The solution F has the Equal Treatment (ET) property if for each $w \in \mathcal{W}^{N'}$ and for each $i, j \in N$ with $C_i(w) = C_j(w)$

$$F_i(w) = F_j(w).$$

Property 3. The solution F has the Upper Bounded Contribution (UBC) property if for each $w \in \mathcal{W}^{N'}$ and every (w, N')-component $C \neq \{0\}$

$$\sum_{i \in C \setminus \{0\}} F_i(w) \leq \min_{i \in C \setminus \{0\}} w(\{i, 0\}).$$

Property 4. The solution F has the Cone-wise Positive Linearity (CPL) property if for each $\sigma \in \Sigma_{E_{N'}}$, for each pair of mcst situations $w, \hat{w} \in K^{\sigma}$ and for each pair $\alpha, \hat{\alpha} \geq 0$, we have

$$F(\alpha w + \hat{\alpha} \hat{w}) = \alpha F(w) + \hat{\alpha} F(\hat{w}).$$

PROPOSITION 2. *The P-value satisfies the properties* EFF, ET, UBC *and* CPL.

Proof. Let $w \in \mathcal{W}^{N'}$ and let $\sigma \in \Sigma_{E_{N'}}$ be such that $w \in K^{\sigma}$. Then the following considerations hold:

(i) Let $\sigma(t_1)$, $\sigma(t_2), \ldots, \sigma(t_n)$, be the n edges of the mcst Γ corresponding to Kruskal order σ. These edges correspond to non-zero columns in M^{σ} and then the sum of elements of each column equals 1. Hence,

$$P(w) = M^{\sigma} w^{\sigma} = \sum_{r=1}^{n} w(\sigma(t_r)) M^{\sigma} e^{t_r},$$

$$\sum_{i \in N} P_i(w) = \sum_{r=1}^{n} w(\sigma(t_r)) \sum_{i \in N} (M^{\sigma} e^{t_r})_i = \sum_{r=1}^{n} w(\sigma(t_r)) = w(\Gamma),$$

which proves the EFF property.

(ii) Note that if w is the zero function then the ET property is trivially satisfied. Consider $w \neq 0$ and define $k = \min\{j | w(\sigma(j)) > 0\}$. Then w^{σ} is of the form $(0, \ldots, 0, w(\sigma(k)), \ldots, w(\sigma(|E_{N'}|)))^t$. Then for each $i \in N$

$$P_i(w) = (M^{\sigma} w^{\sigma})_i = \sum_{r=k}^{|E_{N'}|} (b_i^{\sigma,r-1} - b_i^{\sigma,r}) w(\sigma(r)). \tag{10}$$

Let C be a (w, N')-component and consider two users $i, j \in C$. By Remark 1 this means that i and j are connected in the graph $\langle N', F^{\sigma,k-1} \rangle$ and so also in $\langle N', F^{\sigma,r} \rangle$ for every $r \in \{k, \ldots, |E_{N'}|\}$. Then for each $r \in \{k, \ldots, |E_{N'}|\}$

$$b_i^{\sigma,r-1} - b_i^{\sigma,r} = b_j^{\sigma,r-1} - b_j^{\sigma,r}.$$

Hence, by (10), $P_i(w) = P_j(w)$, which proves the ET property.

(iii) If w is the zero function then the UBC property is trivially satisfied. Consider $w \neq 0$ and let $C \neq \{0\}$ be a (w, N')-component. Note that there exists $m \in \{1, \ldots, |E_{N'}|\}$ such that $\sigma(m) \subseteq C \cup \{0\}$ and $w(\sigma(m)) = \min_{i \in C \setminus \{0\}} w(\{i, 0\})$. Define $k = \min\{j \mid w(\sigma(j)) > 0\}$. If $m < k$, then $0 \in C$ and we are done since nodes in $C \setminus \{0\}$ pay nothing according to $P(w)$. Instead, if $m \geq k$ then

$$\sum_{i\in C\setminus\{0\}} P_i(w) = \sum_{i\in C\setminus\{0\}} \sum_{r=k}^{m} w(\sigma(r))(b_i^{\sigma,r-1} - b_i^{\sigma,r})$$

$$\leq w(\sigma(m)) \sum_{i\in C\setminus\{0\}} \sum_{r=k}^{m} (b_i^{\sigma,r-1} - b_i^{\sigma,r})$$

$$= w(\sigma(m)) \sum_{i\in C\setminus\{0\}} b_i^{\sigma,k-1} = w(\sigma(m)), \qquad (11)$$

where in the first equality we use that $b_i^{\sigma,u} = 0$ for all $u \in \{m, \ldots, |E_{N'}|\}$ and for each $i \in C$, and in the last one we use the fact that all nodes in $C\setminus\{0\}$ are connected in the graph $\langle N', F^{\sigma,k-1}\rangle$. Note that (11) proves the UBC property.

(iv) The CPL property follows trivially from the definition of P. □

THEOREM 1. *The P-value is the unique solution which satisfies the properties* EFF, ET, UBC *and* CPL *on the class* $\mathcal{W}^{N'}$ *of mcst situations.*

Proof. We already know by Proposition 2 that the P-value satisfies the four properties EFF, ET, UBC and CPL. To prove the uniqueness consider a map $\psi : \mathcal{W}^{N'} \to \mathbb{R}^N$ satisfying EFF, ET, UBC and CPL.

Let $\sigma \in \Sigma_{E_{N'}}$ and $k \in \{1, \ldots, |E_{N'}|\}$. First, we will show that for each mcst situation $e^{\sigma,k} \in K^{\sigma}$, $\psi(e^{\sigma,k}) = P(e^{\sigma,k})$. By UBC, for each $(e^{\sigma,k}, N')$-component $C \neq \{0\}$

$$\sum_{i\in C\setminus\{0\}} \psi_i(e^{\sigma,k}) \leq \min_{i\in C\setminus\{0\}} w(\{i,0\}) = \begin{cases} 0 & \text{if } 0 \in C, \\ 1 & \text{if } 0 \notin C, \end{cases} \qquad (12)$$

implying that

$$\sum_{i\in N} \psi_i(e^{\sigma,k}) = \sum_{C\in \mathcal{C}(e^{\sigma,k})} \sum_{j\in C\setminus\{0\}} \psi_j(e^{\sigma,k}) \leq |\mathcal{C}(e^{\sigma,k})| - 1 = e^{\sigma,k}(\Gamma),$$

where Γ is a minimum spanning network on N' for mcst situation $e^{\sigma,k}$. By EFF we have $\sum_{i\in N} \psi_i(e^{\sigma,k}) = e^{\sigma,k}(\Gamma)$, and then inequalities in relation (12) are equalities. Finally, by ET we find that for each $i \in N$

$$\psi_i(e^{\sigma,k}) = \begin{cases} 0 & \text{if } 0 \in C_i(e^{\sigma,k}), \\ \frac{1}{|C_i(e^{\sigma,k})|} & \text{if } 0 \notin C_i(e^{\sigma,k}) \end{cases}$$

$$= \begin{cases} 0 & \text{if } 0 \in C_i(e^{\sigma,k}), \\ \frac{1}{n_i(F^{\sigma,k-1})} & \text{if } 0 \notin C_i(e^{\sigma,k}) \end{cases}$$

$$= P_i(e^{\sigma,k}). \tag{13}$$

Now, we show that for any mcst situation $w \in \mathcal{W}^{N'}$, $\psi(w) = P(w)$. Let $\sigma \in \Sigma_{E_{N'}}$ be such that $w \in K^\sigma$. From the CPL property of ψ and Equation (2) it follows:

$$\psi(w) = w(\sigma(1))\psi(e^{\sigma,1}) + \sum_{k=2}^{|E_{N'}|} (w(\sigma(k)) - w(\sigma(k-1)))\psi(e^{\sigma,k}). \tag{14}$$

Further, from (8), (13) and (14) we obtain $\psi(w) = P(w)$.

To prove the logical independence of the four properties we need to consider some other solutions on $\mathcal{W}^{N'}$:

(i) αP, an α multiple of the solution concept P, with $\alpha \in [0, 1)$;
(ii) ε, such that $\varepsilon_i(w) = w(\Gamma)/|N|$ for $i \in N$, where Γ is a minimum spanning network on N' for mcst situation w;
(iii) P^τ, where $\tau \in \Sigma_N$, the set of bijections on N. To introduce this solution we follow the same plan used for the P-value. Analogously to Definition 3, for each $\sigma \in \Sigma_{E_{N'}}$ we define $P^{\sigma,\tau}(w) = M^{\sigma,\tau} w^\sigma$ for each mcst situation w in the cone K^σ. Similarly to Definition 1, here $M^{\sigma,\tau} e^k = b^{\sigma,\tau,k-1} - b^{\sigma,\tau,k}$, where for each $k \in \{1, \ldots, |E_{N'}| - 1\}$ and $i \in N$, $b^{\sigma,\tau,k}$ is such that

$$b_i^{\sigma,\tau,k} = \begin{cases} 1 & \text{if } i = \arg\min_{j \in C_i(e^{\sigma,k+1})} \tau(j) \text{ and } 0 \notin C_i(e^{\sigma,k+1}), \\ 0 & \text{otherwise,} \end{cases}$$

in addition $b^{\sigma,\tau,0} = 1$ and $b^{\sigma,\tau,|E_{N'}|} = 0$.

A variant of Proposition 1 holds also for the maps $P^{\sigma,\tau}$ with $\sigma \in \Sigma_{E_{N'}}$, and so this enables us to define the solution

$P^{\tau}(w) = P^{\sigma,\tau}(w) = M^{\sigma,\tau} w^{\sigma}$ for each $w \in \mathcal{W}^{N'}$, where $\sigma \in \Sigma_{E_{N'}}$ is such that $w \in K^{\sigma}$. The solution P^{τ} turns out to coincide with the allocation x_N introduced in Norde et al. (2004) via an algorithmic procedure (called the Subtraction Algorithm) for the computation of a population monotonic allocation scheme (pmas) of a mcst game.

(iv) D, such that (w, N')-components "pay" proportionally to their "distance" from the source, i.e. for each $i \in N$

$$D_i(w) = \begin{cases} \frac{1}{|C_i(w)|} \frac{\min_{j \in C_i(w)} w(\{j,0\})}{\sum_{C \in C(w)} \min_{j \in C \setminus \{0\}} w(\{j,0\})} w(\Gamma) & \text{if } 0 \notin C_i(w), \\ 0 & \text{if } 0 \in C_i(w), \end{cases}$$

where Γ is a minimum spanning network on N' for mcst situation w. □

PROPOSITION 3. *The axioms* EFF, ET, UBC *and* CPL *are logically independent.*

Proof. The logical independence of the four properties follows from the following table:

	EFF	ET	UBC	CPL
αP	no	yes	yes	yes
P^{τ}	yes	no	yes	yes
ε	yes	yes	no	yes
D	yes	yes	yes	no

It is left to the reader to check this table. □

5. CONCLUDING REMARKS

In this paper a solution for mcst situations, the *P*-value, has been introduced. Also, an axiomatic characterization of the *P*-value using the properties EFF, ET, UBC and CPL is given. One can prove that

$$P(w) = \frac{1}{n!} \sum_{\tau \in \Sigma_N} P^{\tau}(w), \tag{15}$$

where $w \in \mathcal{W}^{N'}$ and $P^{\tau}(w)$ are as described in Section 4. It is shown in Norde et al. (2004) that $[P_i^{\tau}(w_{|S\cup\{0\}})]_{S\in 2^N\setminus\{\emptyset\},i\in S}$ is a pmas.

The P-value for mcst situations induces a cost sharing rule for mcst games, which we call the Π-value (*Potters value*). The Π-value is the map $\Pi : \mathcal{MCST}^N \rightarrow \mathbb{R}_+^N$ obtained by $\Pi(c_w) = P(w)$, where $w \in \mathcal{W}^{N'}$. It follows that Π is positive linear on the cone $\mathcal{G}^{\sigma}$, i.e. for all $c_w, c_{w'} \in \mathcal{G}^{\sigma}$ and all $\alpha, \alpha' \in \mathbb{R}_+$ it holds

$$\Pi(\alpha c_w + \alpha' c_{w'}) = \alpha\Pi(c_w) + \alpha'\Pi(c_{w'}).$$

Moreover, (15) implies that the Π-value is a population monotonic allocation rule. To be more concrete, let us denote by c_w^S the subgame of c_w with player set S, $S \subseteq N$, defined by $c_w^S(T) = c_w(T)$, for each $T \subseteq S$. The Π-value assigns to each $c_w^S \in \mathcal{MCST}^S$ the P-value of the mcst subsituation $w_{|S'}$, where $S' = S \cup \{0\}$. In formula $\Pi(c_w^S) := P(w_{|S'})$ for each $c_w^S \in \mathcal{MCST}^S$. Then, $[\Pi_i(c_w^S)]_{S\in 2^N\setminus\{\emptyset\},i\in S}$ is a pmas.

In Tijs et al. (2003) we focus on other monotonicity properties of the P-value like cost monotonicity and drop-out monotonicity.

REFERENCES

Aarts, H. (1994), *Minimum Cost Spanning Tree Games and Set Games*, Ph.D. Dissertation, University of Twente, The Netherlands.

Bird, C.G. (1976), On cost allocation for a spanning tree: A game theoretic approach, *Networks* 6, 335–350.

Dutta, B. and Kar, A. (2002), Cost monotonicity, consistency and minimum cost spanning tree games, Warwick Economic Research Paper 629, University of Warwick, UK.

Feltkamp, V. (1995), *Cooperation in Controlled Network Structures*, Ph.D. Dissertation, Tilburg University, The Netherlands.

Feltkamp, V., Tijs, S. and Muto, S. (1994), On the irreducible core and the equal remaining obligations rule of minimum cost spanning extension problems, CentER Discussion Paper 106, Tilburg University, The Netherlands.

Gambarelli, G. and Owen, G. (1994), Indirect control of corporations, *International Journal of Game Theory* 23, 287–302.

Granot, D. and Huberman, G. (1981), On minimum cost spanning tree games, *Mathematical Programming* 21, 1–18.

Kar, A. (2002), Axiomatization of the Shapley value on minimum cost spanning tree games, *Games and Economic Behavior* 38, 265–277.

Kent, K.J. and Skorin-Kapov, D. (1996), *Population Monotonic Cost Allocations on MSTs*, Discussion Paper, State University of New York at Stony Brook, USA.

Kruskal, J.B. (1956), On the shortest spanning subtree of a graph and the traveling salesman problem, in *Proceedings of the American Mathematical Society* 7, 48–50.

Kuipers, J. (1993), On the core of information graph games, *International Journal of Game Theory* 21, 339–350.

Moretti, S., Norde, H., Pham Do, K.H. and Tijs, S. (2002), Connection problems in mountains and monotonic allocation schemes, *Top* 10, 83–99.

Norde, H., Moretti, S. and Tijs, S. (2004), Minimum cost spanning tree games and population monotonic allocation schemes, *European Journal of Operational Research* 154, 84–97.

Owen, G. (1975), On the core of linear production games, *Mathematical Programming* 9, 358–370.

Owen, G. (1995), *Game Theory*, 3rd edn., San Diego: Academic Press.

Owen, G. (1999), *Discrete Mathematics and Game Theory*, Boston: Kluwer Academic Publishers.

Sprumont, Y. (1990), Population monotonic allocation schemes for cooperative games with transferable utility, *Games and Economic Behavior* 2, 378–394.

Tijs, S., Moretti, S., Norde, H. and Branzei, R. (2003), Obligation rules for minimum cost spanning tree situations and their monotonicity properties, Working paper.

Addresses for correspondence: Stefano Moretti, Department of Mathematics, University of Genova, 16146 Genova, Italy and Department of Environmental Epidemiology, National Cancer Research Institute of Genova, Italy. E-mail: moretti@dima.unige.it

Rodica Branzei, Faculty of Computer Science, "Alexandru Ioan Cuza" University, Iasi, Romania. E-mail: branzeir@infoiasi.ro

Henk Norde, CentER and Department of Econometrics and Operations Research, Tilburg University, Tilburg, The Netherlands. E-mail: h.norde@uvt.nl

Stef Tijs, Center and Department of Econometrics and Operations Research, Tilburg University, Tilburg, The Netherlands and Department of Mathematics, University of Genova, Italy. E-mail: S.H.Tijs@uvt.nl

DANIEL GÓMEZ, ENRIQUE GONZÁLEZ-ARANGÜENA, CONRADO MANUEL, GUILLERMO OWEN and MONICA DEL POZO

A UNIFIED APPROACH TO THE MYERSON VALUE AND THE POSITION VALUE

ABSTRACT. We reconsider the Myerson value and the position value for communication situations. In case the underlying game is a unanimity game, we show that each of these values can be computed using the inclusion–exclusion principle. Linearity of both values permits us to calculate them without needing the dividends of the induced games (graph-restricted game and link game). The expression of these dividends is only derived in the existing literature for special communication situations. Moreover, the associated inclusion–exclusion decomposability property depends on what we have called the graph allocation rule. This rule is the relative degree (relative indicator) for the position value (Myerson value).

KEY WORDS: Allocation rules, Myerson value, Position value, Inclusion-exclusion decomposability property

1. INTRODUCTION

A communication situation is a triplet (N, v, L), N being simultaneously the set of players in the coalitional game (N, v) and the set of nodes in the undirected graph (N, L). Behind this definition is the idea of modeling the cooperation under restrictions in the communications: the economic possibilities of every coalition (described by the game) are influenced by the restrictions in the cooperation imposed by the graph. In the seminal work in communication situations, Myerson (1977) introduced the graph-restricted game and characterized its Shapley value (Shapley, 1953), the so-called Myerson value, in terms of efficiency and fairness. Owen (1986) analyzed the relation between the unanimity coefficients of the underlying game and those of the graph-restricted game, showing that this relation is particularly easy in case the graph is a tree. Van den

Theory and Decision **56:** 63–76, 2004.

Nouweland (1993) extended these results to the more general case of cycle-complete graphs.

Borm *et al.* (1992) offered an alternative approach to the study of communication situations. While the graph-restricted game focuses on the role of players-nodes, they proposed to investigate the role of the graph edges. They measured the importance of a link by means of its Shapley value in the so-called link game and they proposed to divide equally this value between its two incident nodes. The total worth obtained in this way for a node-player is known as its position value, that was first introduced in Meessen (1988). Borm et al. (1992) obtained a characterization of this value (and a parallel one for the Myerson value) for communication situations with cycle-free graphs. Finally, they gave a relation between the unanimity coefficients of the link game and those of the underlying game in the same context. Recently, Slikker (2003) provided a characterization of the position value for general communication situations. A coherent overview of theoretical literature on communication situations can be found in Slikker and van den Nouweland (2001).

When calculating the Myerson value or the position value of a communication situation, it is first needed to obtain the corresponding induced game: the graph-restricted game (a point game) for the Myerson value and the link game for the position value. After this, the Shapley value of both games must be computed and, in the case of the position value, it is still necessary to allocate it among the nodes. This is an unmanageable task when the graph contains cycles, even if there are not a lot of connections in the graph. If, on the other hand, we try to use the unanimity coefficients of the induced games to obtain the value, the already existing literature do not help us beyond communication situations with cycle-complete graphs.

Our aim in this paper is to present a unifying approach that allows us to calculate the Myerson value or the position value of a communication situation in a similar way. In our proposal, the induced games vanish without trace. In their place, graph allocation rules appear. These graph allocation rules are not dependent on the game v, and will be very simple functions of the graph.

We will prove that, for communication situations in which v is a unanimity game u_S, the relation between each one of these values and its associated graph allocation rule is given by the well known inclusion–exclusion principle. We have called this property the inclusion–exclusion decomposability.

In the case of the Myerson (position) value, the associated graph allocation rule is the relative indicator (relative degree).

The remainder of the paper is organized as follows. In Section 2 we recall some preliminary definitions and we establish the concept of graph allocation rule. Section 3 is devoted to present the main result in the paper and to illustrate it by means of an example. Some remarks appear in the final Section 4.

2. PRELIMINARIES

2.1. *Miscellany*

An n-person coalitional game is given by a pair (N, v), $N = \{1, \ldots, n\}$ being the set of players and v a real valued function defined on $2^N = \{S/S \subset N\}$, satisfying $v(\emptyset) = 0$. When there is no ambiguity with respect to N, a game (N, v) will be identified with its characteristic function v. We will denote by G^N the vector space of all the games with players set N, which has dimension $2^n–1$, n being the cardinality of N. In this paper, a special basis of G^N, the unanimity basis, consisting of the unanimity games $\{u_S\}_{\emptyset \neq S \subset N}$, will be often used. The characteristic function u_S is defined by

$$u_S(T) = \begin{cases} 1, & S \subset T, \\ 0, & \text{otherwise.} \end{cases}$$

For a given $v \in G^N$, $\{\Delta_v(S)\}_{\emptyset \neq S \subset N}$ is the set of the unanimity coefficients of v (the coordinates of v in the unanimity basis).

Moreover, in G^N it is defined the standard inner product of games:

$$(N, v) \cdot (N, w) = (N, v.w), \text{ where } (v \cdot w)(S) = v(S)w(S), \ \forall S \subset N.$$

In this paper we often deal with G_0^N, the subspace of G^N consisting of the zero normalized games (the games with $v(\{i\}) = 0$ for all $i \in N$). The family $\{u_S\}_{S \subset N, |S| \geq 2}$ is a basis of G_0^N.

A very relevant solution concept for cooperative games is the Shapley value. If $v \in G^N$, the Shapley value of v, $\phi(v)$, is the n-vector with components:

$$\phi_i(v) = \sum_{S \subset N \setminus \{i\}} \frac{(n-s-1)!s!}{n!} [v(S \cup i) - v(S)], i \in N.$$

It is easy to see that: $\phi(u_S) = \delta(N,S)/|S|$, $\delta(N,S)$ being the indicator function of S in N, i.e.: $\delta_i(N,S) = \begin{cases} 1, & \text{if } i \in S, \\ 0, & \text{if } i \in \mathrm{N} \setminus S. \end{cases}$

As ϕ is linear in G^N,

$$\phi(v) = \sum_{\phi \neq S \subset N} \Delta_v(S) \frac{\delta(N,S)}{|S|}.$$

A communication situation is a triplet (N, v, L), where $(N, v) \in G^N$ and (N, L) is a graph. L is a subset of the collection of all unordered pairs $\{i,j\} (i \neq j)$ of elements of N. Each pair $\{i,j\} \in L$ is called an edge or a link. We will denote CS^N the family of all communication situations with players set N and CS_0^N the subfamily of CS^N whose elements are the communication situations with v in G_0^N.

We shall say that a graph (N, L) is connected if it is possible to join any two nodes i and j of N by a sequence of edges from L. A subset S of N is connected in (N, L) if $(S, L|_S)$ is a connected graph, with $L|_S = \{\{i,j\} \in L$ such that $i,j \in S\}$. Given a graph (N, L), $T \subset N$ is a connected component of N (in L) if $(T, L|_T)$ is a connected graph and there is no $T' \neq T$ such that $T \subset T'$ and $(T', L|_{T'})$ is connected. The restrictions in the communications imposed by the graph (N, L) determine a partition N/L of N into its connected components. Similarly, S/L will denote the partition of S into the connected components generated by the graph $(S, L|_S)$.

2.2. *Allocation Rules for Communication Situations: The Myerson Value and the Position Value*

An allocation rule on a class of communication situations is merely a function γ that assigns a vector in R^n to every (N, v, L) in the class. An allocation rule γ is said to be linear if, for every (N, v, L), (N, w, L) in the class and every real numbers α, β:

$$\gamma(N, \alpha v + \beta w, L) = \alpha\gamma(N, v, L) + \beta\gamma(N, w, L).$$

Two outstanding allocation rules for communication situations are the Myerson value and the position value.

Given a communication situation (N, v, L), Myerson (1977) introduced the graph-restricted game (a point game) (N, v^L), which represents the economic possibilities of the players taking into account the available communications. Its characteristic function v^L is defined by

$$v^L(S) = \sum_{T \in S/L} v(T), \quad S \in 2^N. \tag{1}$$

Then the Myerson value of a communication situation (N, v, L), that we will denote by $\mu(N, v, L)$, is merely the Shapley value of the game (N, v^L), i.e., $\mu(N, v, L) = \phi(v^L)$.

Another type of game corresponding to a communication situation was introduced in Meessen (1988). Given (N, v, L) in CS_0^N, the link game $(L, r_L^v) \in G^L$ (the vector space of all games with players set L) has the characteristic function

$$r_L^v(A) = \sum_{T \in N/A} v(T), \quad A \in 2^L.$$

So, a coalition A of links receives the sum of the values of those coalitions of nodes that are connected components of N in A. The position value π is then defined as

$$\pi_i(N, v, L) = \frac{1}{2}\sum_{a \in L_i} \phi_a(r_L^v), \quad i \in N,$$

where L_i is the set of edges incident on node i, i.e., $L_i = \{\{i,j\} \in L : j \in N\}$. Therefore, the position value of a node is the sum of the contributions of the different links incident on that node. The contribution of each edge is obtained equally apportioning the Shapley value of that edge in the link game between its two extreme nodes.

Linearity of the Shapley value guarantees linearity of both the Myerson and the position values.

2.3. *Graph Allocation Rules*

The remainder of this paper depends heavily on the different ways to assign values to the nodes of a graph, independently of the game in the communication situation. Let us note Γ^N the class of all graphs with nodes set N. A graph allocation rule will be merely a function $h : \Gamma^N \to R^n$. Two particular graph allocation rules are:

(a) The relative degree, that is defined as

$$d_i^r(N, L) = \frac{d_i(N, L)}{2|L|}, \quad i \in N \tag{2}$$

$d_i(N, L)$ being the degree of node i in the graph L.

(b) The relative indicator, that is defined as

$$c_i^r(N, L) = \frac{\delta_i(N, D(L))}{|D(L)|}, \quad i \in N \tag{3}$$

$D(L) = \{i \in N \text{ such that } d_i(N, L) > 0\}$ being the set of the not isolated nodes in (N, L).

3. THE UNIFIED APPROACH

In this section we will present our results. They allow computing the Myerson and the position values in a communication situation (with a unanimity game) using the well-known inclusion–exclusion principle. Both values will be expressed as an alternate sum over the connection sets of edges for S. In fact, they differ merely in their associated graph allocation rule. Let us first consider a few definitions.

DEFINITION 1. Given a graph (N, L) and a coalition $S \subset N$ with $|S| \geq 2$, we will say that $L^S \subset L$ is a minimal connection set (of edges) of S in (N, L) if S is connected in (N, L^S) and, for all $A \subset L^S$, $A \neq L^S$, S is not connected in (N, A).

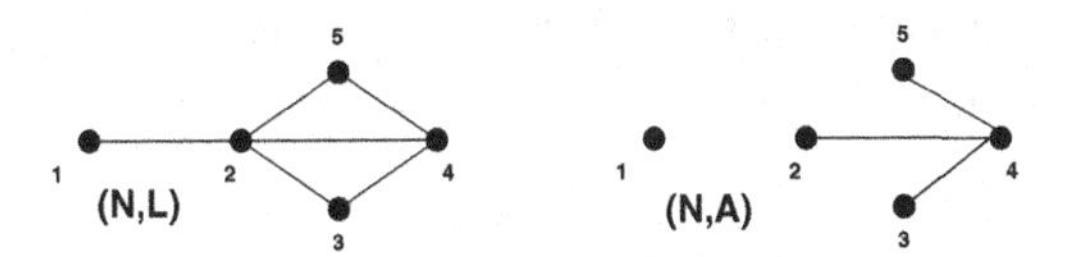

Figure 1. Two graphs.

Let us observe that, given a graph (N, L) and $S \subset N$, several minimal connection sets of S, one or none could exist. We will denote by $MC_L^e(S)$ the collection of these subsets.

EXAMPLE 1. Given (N, L), $N = \{1, 2, 3, 4, 5\}$ and $L = \{\{1,2\}, \{2,3\}, \{2,4\}, \{2,5\}, \{3,4\}, \{4,5\}\}$ (Figure 1), if $S = \{1,4\}$ *then*, $MC_L^e(S) = \{L_1^S, L_2^S, L_3^S\}$, with $L_1^S = \{\{1,2\}, \{2,3\}, \{3,4\}\}$, $L_2^S = \{\{1,2\}, \{2,4\}\}$ and $L_3^S = \{\{1,2\}, \{2,5\}, \{4,5\}\}$.

EXAMPLE 2. Consider (N, A), with $N = \{1, 2, 3, 4, 5\}$ and $A = \{\{2,4\}, \{3,4\}, \{4,5\}\}$ (on the right side in Figure 1). In this case, for $S = \{1,4\}$, we have $MC_A^e(S) = \emptyset$.

DEFINITION 2. An allocation rule $\gamma : CS^N \to R^n$ is inclusion–exclusion decomposable (i.e.d.) in terms of a graph allocation rule h if, for each $(N, u_S, L) \in CS^N$, with $S \subset N$, $\gamma(N, u_S, L)$ satisfies:

(i) $\gamma(N, u_{\{i\}}, L) = \delta(N, \{i\})$, *for all* $i \in N$,

(ii) *If* $|S| \geq 2$ *and* $MC_L^e(S) = \emptyset$, *then* $\gamma(N, u_S, L)$ *is the null vector in* R^n.

(iii) *If* $|S| \geq 2$ *and* $MC_L^e(S) = \{L_k^S\}_{k=1}^{r(S)} \neq \emptyset$ *then*,

$$\gamma(N, u_S, L) = \sum_{k=1}^{r(S)} h(N, L_k^S) - \sum_{k \neq m} h(N, L_k^S \cup L_m^S) + \cdots + (-1)^{r(S)+1} h\left(N, \bigcup_{k=1}^{r(S)} L_k^S\right). \tag{4}$$

In this case, we will also say that γ satisfies the inclusion–exclusion decomposability (i.e.d.) property.

In the particular case of allocation rules $\gamma : CS_0^N \to R^n$, it must be understood that previous definition applies only to communication situations (N, u_S, L) with $|S| \geq 2$.

In order to prove that both the Myerson value and the position value satisfy this property, let us first prove the following lemmas.

LEMMA 1. *Let* $(N, u_S, L) \in CS^N$, *with* $|S| \geq 2$ and $MC_L^e(S) = \{L_k^S\}_{k=1}^{r(S)}$. *Then* (N, u_S^L) *has characteristic function*:

$$u_S^L = \begin{cases} \mathbf{1} - \prod_{k=1}^{r(S)} \left[\mathbf{1} - u_{D(L_k^S)}\right], & \text{if } MC_L^e(S) \neq \emptyset, \\ \mathbf{0}, \text{ if } MC_L^e(S) = \emptyset, \end{cases} \tag{5}$$

where $D(L_k) = \{i \in N : d_i(N, L_k) > 0\}$, *and* $\mathbf{1}$ *being the unit element of the standard inner product in* G^N.

Proof. From (1), the equality holds if $MC_L^e(S) = \emptyset$ as, in this case, $u_S^L = \mathbf{0}$. If $MC_L^e(S) \neq \emptyset$ then, for every $T \subset N$, we have:

$$u_S^L(T) = \begin{cases} 1, & \text{if there exists } K \text{ connected in } (N, L) \\ & \text{such that } S \subset K \subset T, \\ 0, & \text{otherwise.} \end{cases}$$

Furthermore,

$$\left(\mathbf{1} - \prod_{k=1}^{r(S)} \left[\mathbf{1} - u_{D(L_k^S)}\right]\right)(T) = 1$$

holds if and only if there exists some $L_K^S \in MC_L^e(S)$ such that $(\mathbf{1} - u_{D(L_k^S)})(T) = 0$ or, equivalently, $u_{D(L_k^S)}(T) = 1$. But this holds only when $D(L_k^S)$, which is a connected set that contains S, is contained in T. Then equality (5) holds. □

LEMMA 2. *Let* $(N, u_S, L) \in CS_0^N$ *and* $MC_L^e(S) = \{L_k^S\}_{k=1}^{r(S)}$. *Then the link game* $(L, r_L^{u_S})$ *has characteristic function*:

$$r_L^{u_S} = \begin{cases} \mathbf{1} - \prod_{k=1}^{r(S)} \left[\mathbf{1} - u_{L_k^S}\right], & \text{if } MC_L^e(S) \neq \emptyset, \\ \mathbf{0}, \text{ if } MC_L^e(S) = \emptyset. \end{cases} \tag{6}$$

Proof. Suppose S is such that $MC_L^e(S) \neq \emptyset$, as the result is trivial when $MC_L^e(S) = \emptyset$. Then for every $A \subset L$:

$$r_L^{u_S}(A) = \begin{cases} 1, & \text{if there exists } K \text{ connected in } (N, A) \\ & \text{such that } S \subset K, \\ 0, & \text{otherwise.} \end{cases}$$

On the other hand,

$$\left(\mathbf{1} - \prod_{k=1}^{r(S)} \left[\mathbf{1} - u_{L_k^S}\right]\right)(A) = 1,$$

if and only if there exists some $L_k^S \in MC_L^e(S)$ such that $(\mathbf{1} - u_{L_k^S}) \times (A) = 0$, which is equivalent to $L_k^S \subset A$. So $D(L_k^S)$ is connected in (N, A) and such that $S \subset D(L_k^S)$ and then equality (6) holds. □

THEOREM 1. *Both the Myerson value and the position value satisfy i.e.d. property.*

Proof. First, for μ. If $(N, u_S, L) \in CS^N$ and $S = \{i\}$, then $\mu(N, u_{\{i\}}, L) = \varphi(u_{\{i\}}^L) = \varphi(u_{\{i\}}) = \delta(N, \{i\})$. If $|S| \geq 2$ and $MC_L^e(S) = \emptyset$, then $u_S^L = \mathbf{0}$ and the result is trivial. Finally, if $|S| \geq 2$ and $MC_L^e(S) = \{L_k^S\}_{k=1}^{r(S)} \neq \emptyset$ then, by Lemma 1,

$$\mu(N, u_S, L) = \varphi(u_S^L) = \varphi\left(\mathbf{1} - \prod_{k=1}^{r(S)} \left[\mathbf{1} - u_{D(L_k^S)}\right]\right).$$

Moreover, we have

$$\mathbf{1} - \prod_{k=1}^{r(S)} \left[\mathbf{1} - u_{D(L_k^S)}\right]$$

$$= \sum_{k=1}^{r(S)} u_{D(L_k^S)} - \sum_{k \neq m} u_{D(L_k^S) \cup D(L_m^S)} + \cdots + (-1)^{r(S)+1} u_{\bigcup_{k=1}^{r(S)} D(L_k^S)}$$

For every $A_1, A_2 \subset L$, $D(A_1 \cup A_2) = D(A_1) \cup D(A_2)$ holds trivially and thus, using the linearity of the Shapley value, we have

$$\mu(N, u_S, L) = \sum_{k=1}^{r(S)} \varphi(u_{D(L_k^S)}) - \sum_{k \neq m} \varphi\left(u_{D(L_k^S) \cup D(L_m^S)}\right) +$$

$$+ \cdots + (-1)^{r(S)+1} \varphi\left(u_{D\left(\bigcup_{k=1}^{r(S)} L_k^S\right)}\right).$$

Using the Shapley value of the unanimity games and expression in (3), we obtain

$$\mu(N, u_S, L) = \sum_{k=1}^{r(S)} c^r(N, L_k^S) - \sum_{k \neq m} c^r(N, L_k^S \cup L_m^S) +$$

$$+ \cdots + (-1)^{r(S)+1} c^r\left(N, \bigcup_{k=1}^{r(S)} L_k^S\right).$$

Therefore, μ is i.e.d. in terms of the graph allocation rule we have called relative indicator as satisfies (4) with $h = c^r$.

Now, let us consider the position value π. Let $(N, u_S, L) \in CS_0^N$ and $S \subset N$, with $MC_L^e(S) = \{L_k^S\}_{k=1}^{r(S)} \neq \emptyset$ (otherwise the result is trivial). Then we have, for every node $i \in N$:

$$\pi_i(N, u_S, L) = \frac{1}{2} \sum_{a \in L_i} \varphi_a(r_L^{u_S}) = \frac{1}{2} \sum_{a \in L} \delta_a(L, L_i) \varphi_a(r_L^{u_S}), \qquad (7)$$

where, for every $L' \subset L$, $\delta_a(L, L')$ is the indicator function of L' in L, defined as

$$\delta_a(L, L') = \begin{cases} 1, & \text{if } a \in L', \\ 0, & \text{if } a \in L \backslash L'. \end{cases}$$

For a fixed $a \in L$, $\varphi_a(r_L^{u_S}) = \varphi_a(\mathbf{1} - \prod_{k=1}^{r(S)}(\mathbf{1} - u_{L_k^S}))$ by Lemma 2. Moreover we have

$$\mathbf{1} - \prod_{k=1}^{r(S)} (\mathbf{1} - u_{L_k^S}) = \sum_{k=1}^{r(S)} u_{L_k^S} - \sum_{k \neq m} u_{L_k^S \cup L_m^S} + \cdots + (-1)^{r(S)+1} u_{\bigcup_{k=1}^{r(S)} L_k^S},$$

and then, by the linearity of the Shapley value

$$\varphi_a(r_L^{u_S}) = \sum_{k=1}^{r(S)} \varphi_a(u_{L_k^S}) - \sum_{k \neq m} \varphi_a(u_{L_k^S \cup L_m^S}) + \cdots + (-1)^{r(S)+1} \varphi_a\left(u_{\bigcup_{k=1}^{r(S)} L_k^S}\right).$$

Taking into account the expression of the Shapley value for unanimity games:

$$\varphi_a(r_L^{u_S}) = \sum_{k=1}^{r(S)} \frac{\delta_a(L, L_k^S)}{|L_k^S|} - \sum_{k \neq m} \frac{\delta_a(L, L_k^S \cup L_m^S)}{|L_k^S \cup L_m^S|} + + \cdots + (-1)^{r(S)+1} \frac{\delta_a(L, \bigcup_{k=1}^{r(S)} L_k^S)}{|\bigcup_{k=1}^{r(S)} L_k^S|}. \tag{8}$$

And then, by substitution of expression (8) in formula (7), we obtain

$$\begin{aligned}\pi(N, u_S, L) =& \frac{1}{2}\sum_{a \in L} \delta_a(L, L_i) \sum_{k=1}^{r(S)} \frac{\delta_a(L, L_k^S)}{|L_k^S|} - \\ &- \frac{1}{2}\sum_{a \in L} \delta_a(L, L_i) \sum_{k \neq m} \frac{\delta_a(L, L_k^S \cup L_m^S)}{|L_k^S \cup L_m^S|} + \\ &+ \cdots + \frac{1}{2}(-1)^{r(S)+1} \sum_{a \in L} \delta_a(L, L_i) \frac{\delta_a(L, \bigcup_{k=1}^{r(S)} L_k^S)}{|\bigcup_{k=1}^{r(S)} L_k^S|} \\ =& \sum_{k=1}^{r(S)} \frac{1}{2} \sum_{a \in L} \frac{\delta_a(L, L_i \cap L_k^S)}{|L_k^S|} \\ &- \sum_{k \neq m} \frac{1}{2} \sum_{a \in L} \frac{\delta_a(L, L_i \cap (L_k^S \cup L_m^S))}{|L_k^S \cup L_m^S|} + \cdots + \\ &+ (-1)^{r(S)+1} \frac{1}{2} \sum_{a \in L} \frac{\delta_a(L, L_i \cap (\bigcup_{k=1}^{r(S)} L_k^S))}{|\bigcup_{k=1}^{r(S)} L_k^S|},\end{aligned} \tag{9}$$

where, for $k = 1, \ldots, r(S)$, $L_i \cap L_k^S$ is the set of edges of the graph (N, L_k^S) incident on i; $L_i \cap (L_k^S \cup L_m^S)$ is the set of edges of the graph $(N, L_k^S \cup L_m^S)$ incident on i for $k \neq m$, and so on.

Then, taking into account (2), expression (9) is equivalent to

$$\pi_i(N, u_S, L) = \sum_{k=1}^{r(S)} d_i^r(N, L_k^S) - \sum_{k<m}^{r(S)} d_i^r(N, L_k^S \cup L_m^S) + + \cdots + (-1)^{r(S)+1} d_i^r\left(N, \bigcup_{k=1}^{r(S)} L_k^S\right)$$

and therefore the position value is i.e.d. in terms of the relative degree as satisfies expression (4) taking $h = d^r$. □

Trivially from the previous theorem we have:

COROLLARY 1. *The Myerson value is the unique allocation rule on CS^N that satisfies linearity and i.e.d. in terms of the relative indicator graph allocation rule.*

COROLLARY 2. *The position value is the unique allocation rule on CS_0^N that satisfies linearity and i.e.d. in terms of the relative degree graph allocation rule.*

To illustrate the parallel method to calculate the Myerson value and the position value, let us consider the following example.

EXAMPLE 3. Let (N, u_S, L) be a communication situation with (N, L) the graph in Example 1 and $S = \{1, 4\}$. If γ is equal to μ or π:

$$\begin{aligned}\gamma(N, u_{\{1,4\}}, L) = {} & h(N, L_1^S) + h(N, L_2^S) + h(N, L_3^S) - \\ & - h(N, L_1^S \cup L_2^S) - h(N, L_1^S \cup L_3^S) - \\ & - h(N, L_2^S \cup L_3^S) + h(N, L_1^S \cup L_2^S \cup L_3^S)\end{aligned}$$

Now, taking $h = c^r$ we will obtain the Myerson value:

$$\begin{aligned}\mu(N, u_{\{1,4\}}, L) = {} & (1/4,1/4,1/4,1/4,0) + (1/3,1/3,0,1/3,0) + \\ & + (1/4,1/4,0,1/4,1/4) - (1/4,1/4,1/4,1/4,0) - \\ & - (1/5,1/5,1/5,1/5,1/5) - (1/4,1/4,0,1/4,1/4) + \\ & + (1/5,1/5,1/5,1/5,1/5) \\ = {} & (1/3,1/3,0,1/3,0),\end{aligned}$$

and taking $h = d^r$ we will obtain the position value:

$$\begin{aligned}\pi(N, u_{\{1,4\}}, L) = {} & (1/6,1/3,1/3,1/6,0) + (1/4,1/2,0,1/4,0) + \\ & + (1/6,1/3,0,1/6,1/3) - (1/8,3/8,1/4,1/4,0) - \\ & - (1/10,3/10,1/5,1/5,1/5) - (1/8,3/8,0,1/4,1/4) + \\ & + (1/12,1/3,1/6,1/4,1/6) \\ = {} & (19/60,9/20,1/20,2/15,1/20).\end{aligned}$$

4. FINAL REMARKS

(i) From Lemma 1, it is straightforward to obtain the unanimity coefficients of the graph–restricted games u_S^L. Moreover, as a direct consequence of this Lemma 1, every graph-restricted game v^L is in the subspace of G^N spanned by the unanimity games u_S, where S is connected in L. Then Lemma 1 generalizes Theorems 3 and 6 and gives another proof of Theorem 2 in Owen (1986).

(ii) The expression of the unanimity coefficients of the graph-restricted game in terms of the unanimity coefficients of the underlying game (Owen, 1986; Van den Nouweland, 1993) can be easily generalized using Lemma 1.

(iii) From Lemma 2, it is easy to obtain a relation between the unanimity coefficients of the link game and those of the underlying game, that extends to general communication situations the one obtained in Borm *et al.* (1992).

(iv) Even if Corollary 1 (Corollary 2) gives an algebraic characterization of the Myerson (position) value, i.e.d. is not the type of property that can be found appealing in a relevant game theoretical characterization. Perhaps these results should be placed next to the relation between the Myerson (position) value and potentials (see, e.g., Slikker and Van den Nouweland (2001), theorem 4.8).

(v) In Gómez *et al.*(2003) another procedure is given to calculate the Myerson value of a communication situation (N, u_S, L) in terms of the *minimal connection sets* (*of nodes*) of S in (N, L). This method is, in most of the cases, computationally more efficient than the one described here. To illustrate this fact, let us consider again the communication situation in Example 1. For the coalition $S = \{1, 4\}$, we have a unique minimal connection set (of nodes), $N_1^S = \{1, 2, 4\}$. So, directly, $\mu(N, u_S, L) = (1/3, 1/3, 0, 1/3, 0)$. (Compare with the corresponding calculations in Example 3.) It is disappointing that this procedure was not extensible to the case of the position value.

ACKNOWLEDGEMENTS

The authors are grateful to two anonymous referees for their helpful suggestions and comments.

REFERENCES

Borm, P., Owen, G. and Tijs, S. (1992), 'On the position value for communication situations, *SIAM Journal on Discrete Mathematics* 5, 305–320.

Gómez, D., González-Arangüena, E., Manuel, C., Owen, G., Pozo, M. and Tejada, J. (2003), Centrality and Power in social networks: A game theoretic approach, *Mathematical Social Science* 46, 27–54.

Meessen, R. (1988), *Communication Games*, Master's Thesis, Department of Mathematics, University of Nijmegen, The Netherlands (in Dutch).

Myerson, R. (1977), Graph and cooperation in games, *Mathematics of Operations Research* 2, 225–229.

Owen, G. (1986), Values of Graph-restricted games, *SIAM Journal on Algebraic and Discrete Methods* 7, 210–220.

Shapley, L.S. (1953), A value for *n*-person games, in A.W. Tucker and H. Kuhn (eds), *Annals of Mathematics Studies*, Princeton, NJ: Princeton University Press, 307–317.

Slikker, M. (2003), A characterization of the position value, Beta working paper WP-96, Eindhoven, Technische Universiteit Eindhoven, The Netherlands.

Slikker, M. and Van den Nouweland, A. (2001), *Social and Economic Networks in Cooperative Game Theory*, Boston: Kluwer Academic Publishers.

Van den Nouweland, A. (1993), *Games and Graphs in Economic Situations*, Ph.D. Thesis, Tilburg University, Tilburg, The Netherlands.

Addresses for correspondence: Mónica del Pozo, Dept. de Estadística e I.O. III. E.U. de Estadística. Universidad Complutense de Madrid. Av Puerta de Hierro s/n Madrid, 28040, Spain. Fax: +34-913-944-064; E-mail: mpozo@estad.ucm.es

Daniel Gómez, Enrique González-Arangüena, Conrado Manuel, Dept. de Estadística e I.O. III. E.U. de Estadística.Universidad Complutense de Madrid. Av Puerta de Hierro s/n Madrid, 28040, Spain. E-mail: {conrado, dagomez}@estad.ucm.es

Guillermo Owen, Department of Mathematics, Naval Postgraduate School, Monterey, CA 93942, USA. E-mail: gowen@nps.navy.mil

FRANCESC CARRERAS

α-DECISIVENESS IN SIMPLE GAMES

ABSTRACT. We introduce here a generalized decisiveness index for simple games that is able to encompass both normative and strategic features of collective decision-making mechanisms. The mathematical properties of this index and also a related Banzhaf index are studied.

KEY WORDS: decisiveness, assessment, simple game, Banzhaf value

1. INTRODUCTION

In a recent paper (Carreras, 2001), and based on a neutral probabilistic voting model for a proposal P against a given status quo Q, a "structural" decisiveness index has been introduced which applies to any simple game and measures the formal agility of the voting procedures represented by the game. The index takes values between 0 and 1 and, e.g., all decisive games get 1/2. Basic properties of the index and a potential-like relationship to the Banzhaf value (the dummy-independent and non-normalized form introduced by Owen (1975) are stated.

The structural decisiveness index of a game can be calculated in terms of the Owen (1972) multilinear extension (MLE) of the game where it applies, by replacing each variable with 1/2. This suggests that, in fact, any values of the variables might make sense, and this leads to generalized decisiveness and Banzhaf indices.

Thus, by assuming an assessment function α is given, that reflects the inclination of each player to vote for a proposal, we define in Section 2 a *α-decisiveness index*, show that it can be computed by means of the MLE, and describe its first properties. In Section 3, we state the main properties of the index and provide an axiomatic characterization. In Section 4, the potential notion gives rise to a definition of a *Banzhaf α-index*, whose computation in terms of MLEs is also provided by

Theory and Decision **56:** 77–91, 2004.

generalizing Owen's (1975) result for the classical Banzhaf value, and we give its main properties.

The new decisiveness and Banzhaf indices will be helpful to analyze also strategic voting aspects that the formal ones, rather intended for normative questions only, cannot cope with. Some examples are given in Section 5.

2. THE α-DECISIVENESS INDEX

Let us first merely recall that a (*monotonic*) *simple game* is a pair (N, W) where N is a finite set of *players* and W is a collection of *coalitions* (subsets of N) that satisfies the following properties: (1) $\emptyset \notin W$; (2) if $S \in W$ and $S \subset T \subseteq N$ then $T \in W$ (*monotonicity*). A coalition S is *winning* if $S \in W$, and *losing* otherwise. Due to monotonicity, the set W^m of *minimal* winning coalitions determines the game. If $n = |N|$ we will usually take $N = \{1, 2, \ldots, n\}$. We will denote as SG the set of all simple games.

DEFINITION 2.1. Given N, any function $\alpha : N \to [0, 1]$ will be called an *assessment* on N. Equivalently, α may be thought of as a vector $\alpha = (\alpha_1, \alpha_2, \ldots, \alpha_n) \in [0, 1]^N$. By ASG we will denote the set of all *assessed simple games*, i.e. triples (N, W, α) where (N, W) is a simple game and α is an assessment on N.

DEFINITION 2.2. The *(generalized) decisiveness index* is the map $\delta : ASG \to R$ (set of real numbers) given by

$$\delta(N, W, \alpha) = \sum_{S \in W} \prod_{i \in S} \alpha_i \prod_{j \in N \setminus S} (1 - \alpha_j) \text{ for every } (N, W, \alpha).$$

Number $\delta(N, W, \alpha)$ will be called the α-decisiveness index of game (N, W). By introducing

$$\pi_S(\alpha) = \prod_{i \in S} \alpha_i \prod_{j \in N \setminus S} (1 - \alpha_j)$$

for all $S \subseteq N$ and all α on N, in order to alleviate the notation at some points below, we have $\delta(N, W, \alpha) = \sum_{S \in W} \pi_S(\alpha)$.

Remark 2.3. (a) Let us assume that a proposal P to modify a given status quo Q has been submitted to the members of N. The proposal will pass if and only if the set of members that vote for P is a winning coalition $S \in W$, so that abstention is allowed but it counts for Q. If each agent $i \in N$ has an independent probability α_i to vote for P, and hence a probability $1 - \alpha_i$ to vote for Q or abstain, then $\delta(N, W, \alpha)$ is easily seen to be the probability of the proposal to be socially accepted by N.

(b) Assessment α is thus implicitly assumed to reflect the inclination of each member of N towards a given proposal P. Then, in general, it will depend on all and each one of the members of N, on P and even on Q, but we will suppose that the inclination of each $i \in N$ towards Q is, simply, $1 - \alpha_i$. Instead, α will not be assumed to depend on W, i.e., on the rules that state the approval of a proposal, although this possibility might well deserve further research.

(c) An interesting framework is obtained when each agent can be located in a classical left-to-right ideological axis where, to fix ideas, we take $\alpha_i = 0$ as the most conservative position and $\alpha_i = 1$ as the most progressive one and assume, coherently, that the status quo Q is conservative and proposal P is progressive. A second setup, of interest in countries where a strong nationalist spirit exists (like e.g. Catalonia or the Basque Country), arises by similarly opposing nationalism to centralism (referred to the state to which the country belongs).

(d) By setting $\alpha_i = 1/2$ for all $i \in N$ we get Carreras' (2001) *structural decisiveness index*, defined as a map $\delta : SG \rightarrow R$ given by

$$\delta(N, W) = 2^{-n}|W| \text{ for all } (N, W)$$

and assumed to provide a measure of the *formal* effectiveness *in abstracto* of the game as reflecting a mechanism to take decisions, that is, without taking into account either the personality of concrete agents or their preferences as to a concrete proposal P versus a concrete status quo Q.

Remark 2.4. It is obvious that if $f(x_1, x_2, \ldots, x_n)$ is the MLE (Owen, 1972, 1995) of game (N, W) then $\delta(N, W, \alpha) = f(\alpha_1, \alpha_2, \ldots, \alpha_n)$ and, in particular, $\delta(N, W) = f(1/2, 1/2, \ldots,$

1/2). For example, the MLE of a unanimity game (N, U_S), where $S \subseteq N$, is $f(x_1, x_2, \ldots, x_n) = \prod_{i \in S} x_i$, so that

$$\delta(N, U_S, \alpha) = \prod_{i \in S} \alpha_i \text{ for all } \alpha.$$

It follows that each value $f(\alpha_1, \alpha_2, \ldots, \alpha_n)$ of the MLE of a simple game can be interpreted as the probability of a proposal to pass in our above voting model when each agent i has a probability α_i to vote for, and this reveals the meaning of the MLE of a simple game in a new light. From the probabilistic interpretation of $\delta(N, W, \alpha)$, it follows that $0 \leq \delta(N, W, \alpha) \leq 1$ for all (N, W, α).

PROPOSITION 2.5. *Let (N,W) be a non-empty game, L be the set of its losing coalitions and (N,W^*) be the dual game of (N, W). Then, for all α:*

(a) $\delta(N, W, \alpha) + \delta(N, W^*, 1 - \alpha) = 1,$

(b) $$\delta(N, W, \alpha) = \frac{1}{2} + \frac{1}{2}\left[\sum_{S \in W} \pi_S(\alpha) - \sum_{S \in L} \pi_S(\alpha)\right].$$

Proof. (a) The statement easily follows from the fact that the MLEs f and f^*, of games (N, W) and (N, W^*), respectively, satisfy the relationship

$$f(x) + f^*(1 - x) = 1 \quad \text{for all } x \in [0, 1]^n,$$

already given in Owen (1972, 1995) in the more general context of cooperative games but only for constant-sum games.

(b) In a similar way, one checks that

$$\delta(N, W, \alpha) - \delta(N, W^*, 1 - \alpha) = \sum_{S \in W} \pi_S(\alpha) - \sum_{S \in L} \pi_S(\alpha).$$

Then it suffices to add term by term this and the formula given in (a). □

3. MAIN PROPERTIES AND AXIOMATIC CHARACTERIZATION

Our goal in this section is to analyze the behavior of the α-decisiveness index with regard to the basic forms of compounding simple games and to derive an axiomatic characterization for this measure.

THEOREM 3.1 *Let* (N, W, α) *be an assessed simple game.*

(a) *If* (M, W^M) *is the null extension of* (N, W) *to a superset* $M \supset N$, *given by* $W^M = \{S \subseteq M : S \in W\}$, *and* α^M *is any extension of* α *to* M, *then* $\delta(M, W^M, \alpha^M) = \delta(N, W, \alpha)$.

(b) *Let* $S \subset N$ *and* $(N_{-S}, W_{-S}, \alpha_{-S})$ *be the assessed subgame that arises when all players of* S *leave the game, where* $N_{-S} = N \setminus S$, $W_{-S} = \{T \subseteq N \setminus S : T \in W\}$ *and* α_{-S} *is the restriction of* α to $N \setminus S$. *Then*

$$\delta(N_{-S}, W_{-S}, \alpha_{-S}) \leq \delta(N, W, \alpha),$$

and the equality holds iff for each $i \in S$ *either* $\alpha_i = 0$ *or* i *is a null player in* (N, W).

(c) *If* $(N, W, \alpha) = (N_1, W_1, \alpha_1) \times (N_2, W_2, \alpha_2) \times \cdots \times (N_r, W_r, \alpha_r)$, *in the sense that* $\{N_1, N_2, \ldots, N_r\}$ *is a partition of* N, $S \in W$ *iff* $S \cap N_i \in W_i$ *for* $i = 1, 2, \ldots, r$, *and* α_i *is the restriction of* α *to* N_i *for* $i = 1, 2, \ldots, r$, *then*

$$\delta(N, W, \alpha) = \prod_{i=1}^{r} \delta(N_i, W_i, \alpha_i).$$

(d) *If* (N, W', α) *is another assessed simple game, then*

$$\delta(N, W \cup W', \alpha) \\ = \delta(N, W, \alpha) + \delta(N, W', \alpha) - \delta(N, W \cap W', \alpha).$$

(e) *If, moreover, the sets* E *and* E' *of non-null players, of* (N, W) *and* (N, W') *respectively, are disjoint, then*

$$\delta(N, W \cap W', \alpha) = \delta(N, W, \alpha)\delta(N, W', \alpha).$$

Proof. (a) It suffices to consider the case where $|M \setminus N| = 1$, and we will assume, without loss of generality, $M = N \cup \{n+1\}$. Let $f(x_1, x_2, \ldots, x_n)$ and $g(x_1, x_2, \ldots, x_n, x_{n+1})$ be the MLEs of (N, W) and (M, W^M), respectively. As $n+1$ is a null player in (M, W^M) and $W = (W^M)_{-\{n+1\}}$ (residual subgame when $n+1$ leaves), g is independent of x_{n+1}, f can be obtained from g by replacing x_{n+1} with 0 (elementary properties of the MLEs) and therefore

$$g(x_1,x_2,\dots,x_n,x_{n+1}) = g(x_1,x_2,\dots,x_n,0) = f(x_1,x_2,\dots,x_n),$$

whence

$$\delta(M,W^M,\alpha^M) = g(\alpha_1^M,\dots,\alpha_n^M,\alpha_{n+1}^M) = f(\alpha_1,\alpha_2,\dots,\alpha_n) = \delta(N,W,\alpha).$$

(b) Again it suffices to deal with a particular case, where $|N\backslash S| = 1$, and we can assume $S = \{n\}$. Then, by using again the elementary property of the MLE with regard to subgames,

$$\delta(N,W,\alpha) - \delta(N_{-\{n\}},W_{-\{n\}},\alpha_{-\{n\}}) = \alpha_n \sum_{\substack{S\in W\\ n\in S\\ S\backslash\{n\}\notin W}} \prod_{\substack{i\in S\\ i\neq n}} \alpha_i \prod_{j\in N\backslash S} (1-\alpha_j) \geq 0.$$

And, obviously, this difference vanishes iff either $\alpha_n = 0$ or n is a null player in (N, W).

(c) As (N, W) is the composition of games (N_1, W_1), $(N_2, W_2), \dots, (N_r, W_r)$ by means of the r-person unanimity game played by the set $R = \{N_1,\dots,N_r\}$, that is, $v = u_R[v_1, v_2,\dots, v_r]$ in classical characteristic function notation, it follows (see Owen, 1972, 1995) that the MLE of (N, W) is the product of the MLEs of the compounding games. Using a suitable notation for players,

$$f(x_1^1,\dots,x_{n_1}^1,\dots,x_1^r,\dots,x_{n_r}^r) = f_1(x_1^1,\dots,x_{n_1}^1)\dots f_r(x_1^r,\dots,x_{n_r}^r),$$

and therefore it readily follows that

$$\delta(N, W,\alpha) = \prod_{i=1}^{r} \delta(N_i, W_i, \alpha_i).$$

(d) Since $\{W\backslash W', W'\backslash W, W\cap W'\}$ is a partition of $W\cup W'$,

$$\begin{aligned}\delta(N,W\cup W',\alpha) &= \sum_{S\in W\cup W'} \pi_S(\alpha)\\ &= \sum_{S\in W}\pi_S(\alpha) + \sum_{S\in W'}\pi_S(\alpha) - \sum_{S\in W\cap W'}\pi_S(\alpha)\\ &= \delta(N,W,\alpha) + \delta(N,W',\alpha) - \delta(N,W\cap W',\alpha).\end{aligned}$$

(e) Under the hypotheses made, $(N, W\cap W')$ is isomorphic to the null extension to N of the subgame product $(E, W_E)\times(E', W_{E'})$, where e.g. $W_E = \{S\subseteq E : S\in W\}$. Let α_E and $\alpha_{E'}$ be the restrictions of α to E and E', respectively. By applying (a) and (c) we obtain

$$\delta(N, W \cap W', \alpha) = \delta(E, W_E, \alpha_E)\delta(E', W_{E'}, \alpha_{E'})$$
$$= \delta(N, W, \alpha)\delta(N, W', \alpha). \qquad \square$$

In order to obtain an axiomatic characterization of the α-decisiveness index, we shall consider the following properties:

(*A1*) *Transfer property*. For all (N, W_1), (N, W_2) and α on N,

$$\delta(N, W_1 \cup W_2, \alpha) = \delta(N, W_1, \alpha) + \delta(N, W_2, \alpha) - \delta(N, W_1 \cap W_2, \alpha).$$

This means that the aggregate α-decisiveness arising from games (N, W_1) and (N, W_2) is exactly transferred to (i.e. shared among) games $(N, W_1 \cup W_2)$ and $(N, W_1 \cap W_2)$.

(*A2*) *Null player property*. If $i \notin N$, $M = N \cup \{i\}$ and α^M is any extension to M of a given assessment α on N, then $\delta(M, W^M, \alpha^M) = \delta(N, W, \alpha)$ for all (N, W). Neither the adjunction nor the suppression of one or more null players will affect the α-decisiveness of any game, whichever are the particular assessments of these players.

(*A3*) *Unanimity property*. $\delta(N, U_N, \alpha) = \prod_{i \in N} \alpha_i$ for all N and all α on N. This measures the α-decisiveness of unanimity games.

The independence of the axiomatic system is clear. And, for not to include a trivial axiom such as $\delta(N, \emptyset, \alpha) = 0$ for all α, we shall restrict ourselves to consider the set ASG^+ of all *nonempty* ($W \neq \emptyset$) assessed simple games.

THEOREM 3.2. *A function* $\delta : ASG^+ \to R$ *satisfies properties* A1, A2 *and* A3 *iff it is (the restriction of) the α-decisiveness index.*

Proof. (a) (Existence) As is shown in Theorem 3.1(a) and (d) and Remark 2.4, the α-decisiveness index satisfies A1, A2 and A3.

(b) (Uniqueness) Let δ' be a function satisfying A1, A2 and A3. From A1 and using recurrence it follows that

$$\delta'(N, W_1, \cdots, W_r, \alpha) = \sum_{j=1}^{r} (-1)^{j+1} \sum_{1 \leq i_1 < \cdots < i_j \leq r} \delta'(N, W_{i_1} \cap \cdots \cap W_{i_j}, \alpha).$$

In addition, if (N, W) is a game and $W^m = \{S_1, S_2, \cdots, S_r\}$ then

$$W = U_{S_1} \cup U_{S_2} \cup \cdots \cup U_{S_r}.$$

Moreover,

$$U_{S_{i_1}} \cap U_{S_{i_2}} \cap \cdots \cap U_{S_{i_k}} = U_{S_{i_1} \cup S_{i_2} \cup \cdots \cup S_{i_k}}.$$

Then, by A3, δ' coincides with δ on all assessed unanimity games of the form (N, U_N, α). By A2, they also coincide on any assessed unanimity game of the form (N, U_S, α) with $S \subset N$. Thus, from the formulas derived before we conclude that $\delta' = \delta$ on ASG^+. □

4. THE BANZHAF α-INDEX

We contend that, in some manner, the Banzhaf value measures decisiveness from a *local* viewpoint, i.e. from the perspective of each agent. Then we will modify here the classical Banzhaf value in order to adapt it to the existence of an assessment.

Based on the potential notion introduced by Hart and Mas-Colell (1989) for the Shapley value and extended to the Banzhaf value by Dragan (1996) (see also Dragan (1999) and Calvo and Santos (1997)), in Carreras (2001) it is shown that twice the structural decisiveness index is the unique potential function for the Banzhaf value β on simple games: essentially this means that, if (N, W) is a simple game and $i \in N$, then $\beta_i(N, W) = 2\delta(N, W) - 2\delta(N_{-\{i\}}, W_{-\{i\}})$. Notice that this equation could be taken as a *definition* of the classical Banzhaf value – in terms of structural decisiveness.

Therefore, a natural extension to assessed simple games is provided by the following definition. Unfortunately (but not too much), in the sequel we have to restrict ourselves to deal with strictly positive assessments only, that is, in this section α will be such that $\alpha_i > 0$ for all $i \in N$. Let us denote by ASG^{++} the set of all *positively* ($\alpha > 0$) assessed simple games.

DEFINITION 4.1. The *generalized Banzhaf index* θ is the correspondence that assigns to every $(N, W, \alpha) \in ASG^{++}$ a vector $\theta(N, W, \alpha) \in R^N$ defined by

$$\theta_i(N, W, \alpha) = \frac{1}{\alpha_i}\left[\delta(N, W, \alpha) - \delta(N_{-\{i\}}, W_{-\{i\}}, \alpha_{-\{i\}})\right] \text{ for each } i \in N.$$

Vector $\theta(N, W, \alpha)$ will be called the *Banzhaf α-index* of game (N, W). Notice that we will always have $\theta(N, W, \alpha) \geq 0$.

PROPOSITION 4.2. *Let f be the MLE of game (N, W). Then, for all $\alpha > 0$ on N,*

$$\theta_i(N, W, \alpha) = \frac{\partial f}{\partial x_i}(\alpha_1, \alpha_2, \ldots, \alpha_n) \textit{ for each } i \in N.$$

Proof. By Definitions 2.2 and 4.1, and using the mean value theorem,

$$\theta_i(N, W, \alpha) = \frac{1}{\alpha_i}[f(\ldots, \alpha_i, \ldots) - f(\ldots, 0, \ldots)] = \frac{\partial f}{\partial x_i}(\ldots, \gamma, \ldots)$$

for some $\gamma \in (0, \alpha_i)$. However, $\partial f / \partial x_i$ does not depend on x_i since f is a multilinear function. Then

$$\theta_i(N, W, \alpha) = \frac{\partial f}{\partial x_i}(\alpha_1, \alpha_2, \ldots, \alpha_i, \ldots, \alpha_n). \qquad \square$$

DEFINITION 4.3. Let us call $\pi(N, W, \alpha) = \sum_{i \in N} \theta_i(N, W, \alpha)$ the total α-power of game (N, W). Notice that

$$\pi(N, W, \alpha) = \sum_{i \in N} \frac{\partial f}{\partial x_i}(\alpha_1, \alpha_2, \ldots, \alpha_n),$$

where f is the MLE of game (N, W) (*total α-power property*).

From Definition 4.1, we get

$$\delta(N, W, \alpha) = \frac{1}{\sum_{i \in N} \frac{1}{\alpha_i}} \left[\pi(N, W, \alpha) + \sum_{i \in N} \frac{\delta(N_{-\{i\}}, W_{-\{i\}}, \alpha_{-\{i\}})}{\alpha_i} \right]$$

that is the generalization of Dragan's (1996) *recursive formula*.

Given a function $Q : ASG^{++} \to R$, we define *marginal contributions* to Q in each assessed simple game (N, W, α) by

$$D_i Q(N, W, \alpha) = \frac{1}{\alpha_i}[Q(N, W, \alpha) - Q(N_{-\{i\}}, W_{-\{i\}}, \alpha_{-\{i\}})] \text{ for every } i \in N.$$

We will say that Q is *standard for one-person assessed simple games* if $Q(\{i\}, W, \alpha) = \alpha_i$ if $W = \{\{i\}\}$ and vanishes if $W = \emptyset$.

DEFINITION 4.4. We will say that Q is a *potential function* for θ if

(1) Q is standard for one-person assessed simple games, and
(2) $\sum_{i \in N} D_i Q(N, W, \alpha) = \pi(N, W, \alpha)$.

THEOREM 4.5. *The restriction of δ to ASG^{++} is the unique potential function for θ. Moreover, $D_i\delta(N, W, \alpha) = \theta_i(N, W, \alpha)$ for all (N, W, α) and all $i \in N$.*

Proof. (a) (Existence) It is easily checked that the restriction of δ to ASG^{++} satisfies (1). As to (2), it follows from the fact that $D_i\delta(N, W, \alpha) = \theta_i(N, W, \alpha)$ and the total α-power definition.

(b) (Uniqueness) Let Q satisfy (1) and (2). We use induction on $n = |N|$ to show that $Q(N, W, \alpha) = \delta(N, W, \alpha)$ for all (N, W, α). If $n = 1$ the equality follows at once. If $n > 1$, from (2), Dragan's generalized recursive formula (also applied to Q), and the inductive hypothesis, we get $Q = \delta$ on every (N, W, α). □

We collect in the following statement a series of properties satisfied by the Banzhaf α-index and provide, next, an axiomatic characterization for this index.

THEOREM 4.6. *Let $(N, W, \alpha$ be a positively assessed simple game).*

(a) (*Null player property*) *If i is a null player in (N, W), $\theta_i(N, W, \alpha) = 0$.*
(b) (*Symmetry*) *If $i, j \in N$ are substitutes of each other in (N, W) and $\alpha_i = \alpha_j$, then $\theta_i(N, W, \alpha) = \theta_j(N, W, \alpha)$.*
(c) (*Balanced contributions*) *For all $i, j \in N$,*

$$\alpha_i\left[\theta_i(N, W, \alpha) - \theta_i(N_{-\{j\}}, W_{-\{j\}}, \alpha_{-\{j\}})\right] = \alpha_j\left[\theta_j(N, W, \alpha) - \theta_j(N_{-\{i\}}, W_{-\{i\}}, \alpha_{-\{i\}})\right].$$

(d) *(Duality) If (N, W) is nonempty, (N, W^*) is the dual game of (N, W) and $\alpha < 1$, then $\theta(N, W^*, 1 - \alpha) = \theta(N, W, \alpha)$.*
(e) *(Transfer property) If (N, W', α) is another positively assessed simple game, then*

$$\theta(N, W \cup W', \alpha) = \theta(N, W, \alpha) + \theta(N, W', \alpha) - \theta(N, W \cap W', \alpha).$$

Proof. (a) It is a consequence of the fact that the MLE of (N, W) is a function f independent of x_i and hence its derivative vanishes.

(b) As x_i and x_j appear symmetrically in the expression of the MLE of game (N, W), the result follows immediately from $\alpha_i = \alpha_j$.

(c) It suffices to notice that

$$\alpha_i\big[\theta_i(N, W, \alpha) - \theta_i(N_{-\{j\}}, W_{-\{j\}}, \alpha_{-\{j\}})\big]$$
$$= \delta(N, W, \alpha) - \delta(N_{-\{i\}}, W_{-\{i\}}, \alpha_{-\{i\}})$$
$$- \delta(N_{-\{j\}}, W_{-\{j\}}, \alpha_{-\{j\}}) + \delta(N_{-\{i,j\}}, W_{-\{i,j\}}, \alpha_{-\{i,j\}})$$

is a symmetrical expression with regard to i and j.

(d) It easily follows from the relationship between the MLEs of both games (see the proof of Proposition 2.5).

(e) The proof is straightforward taking into account that, for each $i \in N$,

$$(W \cup W')_{-\{i\}} = W_{-\{i\}} \cup W'_{-\{i\}} \text{ and } (W \cap W')_{-\{i\}} = W_{-\{i\}} \cap W'_{-\{i\}}.$$

Let us consider the following properties for a function θ defined on ASG^{++} that assigns to every assessed simple game (N, W, α) a vector $\theta(N, W, \alpha) \in R^N$.

($B1$) *Transfer property*. For all (N, W_1) and (N, W_2) and any assessment $\alpha > 0$ on N,

$$\theta(N, W_1 \cup W_2, \alpha) = \theta(N, W_1, \alpha) + \theta(N, W_2, \alpha) - \theta(N, W_1 \cap W_2, \alpha).$$

($B2$) *Null player property*. If $i \in N$ is a null player in (N, W) then $\theta_i(N, W, \alpha) = 0$ for any assessment $\alpha > 0$ on N.

($B3$) *Unanimity property*. For all N and any assessment $\alpha > 0$ on N,

$$\theta_i(N, U_N, \alpha) = \alpha_i^{-1} \prod_{j \in N} \alpha_j \text{ for all } i \in N. \qquad \square$$

THEOREM 4.7. *A function θ on ASG^{++} as above satisfies properties B1, B2 and B3 iff it is the Banzhaf α-index.*

Proof. The proof is analogous to that of Theorem 3.2. □

5. EXAMPLES

EXAMPLE 5.1. The structural decisiveness for all 4-person games is provided in Carreras (2001). It is worthy of mention that the decisiveness index can strongly change when dropping the assumption $\alpha_i = 1/2$ for all $i \in N$. For instance:

(a) The MLE of game $(N, W) \equiv [3; 2, 1, 1, 1]$, a decisive weighted majority game, is

$$f(x_1, x_2, x_3, x_4) = x_1(x_2 + x_3 + x_4 - x_2x_3 - x_2x_4 - x_3x_4) + x_2x_3x_4.$$

Let $\alpha = (\alpha_1, 1/2, 1/2, 1/2)$ and α_1 range from 0 to 1. Then $\delta(N, W, \alpha)$ ranges from 0.125 to 0.875 (up to a 75% of variation with respect to structural decisiveness, that is $\delta(N, W) = 1/2$). Instead, the Banzhaf index remains constant: $\theta_1(N, W, \alpha) = 0.75$ and $\theta_i(N, W, \alpha) = 0.25$ for $i = 2, 3, 4$ and all $\alpha_1 \in [0, 1]$.

(b) The MLE of game $(N, W) \equiv [5; 3, 2, 1, 1]$, a proper and weak game where player 1 enjoys veto right, is

$$f(x_1, x_2, x_3, x_4) = x_1(x_2 + x_3x_4 - x_2x_3x_4).$$

If α is as in (a), $\delta(N, W, \alpha)$ ranges from 0 to 0.625 (a 100% of variation with regard to $\delta(N, W) = 0.3125$). Here the Banzhaf index changes with α_1, since $\theta_1(N, W, \alpha) = 0.625$ for all α_1 but $\theta_2(N, W, \alpha)$ ranges from 0 to 0.75 and $\theta_3(N, W, \alpha) = \theta_4(N, W, \alpha)$ ranges from 0 to 0.25.

EXAMPLE 5.2 (Many versus few players). The structural decisiveness of *any* decisive game is 1/2. However, intuitively one might feel that a decisive game with "many" players should be less decisive than another with "few" players, because it seems more difficult to get an agreement in the former case. This is only partially true, as we will see by applying the (generalized) decisiveness index to a numerical instance.

Let, e.g., $(N, W) \equiv [3; 1, 1, 1, 1, 1]$ and $(N', W') \equiv [5; 1, 1, 1, 1, 1, 1, 1, 1, 1]$. As they are decisive games, $\delta(N, W) = 0.5 = \delta(N', W')$. But this implicitly corresponds to assessments α and α' such that $\alpha_i = \alpha'_j = 0.5$ for all $i \in N$ and $j \in N'$. Instead:

(a) If $\alpha_i = \alpha'_j = 0.4$ for all $i \in N$ and $j \in N'$, then $\delta(N, W, \alpha) = 0.3174$ whereas $\delta(N', W', \alpha') = 0.2667$.
(b) If $\alpha_i = \alpha'_j = 0.7$ for all $i \in N$ and $j \in N'$, then $\delta(N, W, \alpha) = 0.8369$ whereas $\delta(N', W', \alpha') = 0.9012$.

EXAMPLE 5.3 (The "Congreso de los Diputados", 1993–1996). The party structure of the Spanish Parliament Lower House during the legislature started in 1993 can be described by the weighted majority game

$$(N, W) \equiv [176; 159, 141, 18, 17, 5, 4, 2, 1, 1, 1, 1]$$

where $W^m = \{\{1,2\},\{1,3\},\{1,4\},\{2,3,4\}\}$. As it is a decisive game its structural decisiveness is $\delta(N, W) = 0.5$, and the power distribution is given by

$$\beta(N, W) = (0.75, 0.25, 0.25, 0.25, 0, \ldots, 0).$$

Let us introduce an assessment according to the guidelines suggested in Remark 2.3(c). The only nonnull players are the four main parties, namely:

- PSOE *(Partido Socialista Obrero Español)*, moderate left-wing party.
- *PP (Partido Popular)*, conservative party.
- *IU (Izquierda Unida)*, coalition of small communist parties.
- CiU *(Convergència i Unió)*, Catalan nationalist middle-of-the-road coalition of two federated parties.

We will first take $\alpha_1 = 0.6$, $\alpha_2 = 0.3$, $\alpha_3 = 0.8$ and $\alpha_4 = 0.5$ (in view of Theorems 3.1(b) and 4.6(a), there is no need to specify α_i for $i \geq 5$). Then the decisiveness of the House (say, its effectiveness to pass a progressive motion) becomes $\delta(N, W, \alpha) = 0.606$. The relevance of each main party to this end in this so assessed situation is given by $\theta_1(N, W, \alpha) = 0.81$, $\theta_2(N, W, \alpha) = 0.22$, $\theta_3(N, W, \alpha) = 0.27$ and $\theta_4(N, W, \alpha) = 0.18$, where the influence of assessment α is evident.

Notice that the "absolute" power of each party does not depend on its own ideological position, but only on the others' one. However, this assertion is somewhat fictitious because moving one's position will do imply variations in the Banzhaf α-index of the remaining players and hence in one's "relative" power. Let us check this by modifying party 2's ideological position in both directions. Table 1 shows several situations. The reader is invited to analyze this table by him/herself and especially look at the percentages of power that appear in each case in the secondary row.

TABLE I
Changes in the Congreso de los Diputados when PP moves ideologically

Assessment (α)	Decisiveness (δ)	Power distribution (θ and %)	Total power (π)
(0.6,0.5,0.8,0.5)	0.650	(0.75,0.22,0.25,0.22) (52.08,15.28,17.36,15.28)	1.44
(0.6,0.4,0.8,0.5)	0.628	(0.78,0.22,0.26,0.20) (53.42,15.07,17.81,13.70)	1.46
(0.6,0.3,0.8,0.5)	0.606	(0.81,0.22,0.27,0.18) (54.73,14.86,18.24,12.16)	1.48
(0.6,0.2,0.8,0.5)	0.584	(0.84,0.22,0.28,0.16) (56.00,14.67,18.67,10.67)	1.50
(0.6,0.1,0.8,0.5)	0.562	(0.87,0.22,0.29,0.14) (57.24,14.47,19.08,9.21)	1.52

ACKNOWLEDGEMENT

Research partially supported by Grant BFM 2003-01314 of the Science and Technology Spanish Ministry and the European Regional Development Fund.

REFERENCES

Calvo, E. and Santos, J.C. (1997), Potentials in cooperative TU-games, *Mathematical Social Sciences* 34, 175–190.

Carreras, F. (2001), A decisiveness index for voting and other systems, Research Report MA2-IR-01-00007 of the Polytechnic University of Catalonia. Also forthcoming in *European Journal of Operational Research*.

Dragan, I. (1996), New mathematical properties of the Banzhaf value, *European Journal of Operational Research* 95, pp. 451–463.

Dragan, I. (1999), Potentials and consistency for semivalues of finite cooperative TU games, *International Journal of Mathematics, Game Theory and Algebra* 9, pp. 85–97.

Hart, S. and Mas-Colell, A. (1989), Potential, value and consistency, *Econometrica* 57, 589–614.

Owen, G. (1972), Multilinear extensions of games, *Management Science* 18, 64–79.

Owen, G. (1975), Multilinear extensions and the Banzhaf value, *Naval Research Logistics Quarterly* 22, 741–750.
Owen, G. (1995), *Game Theory*, 3rd edn. New York: Academic Press.

Address for correspondence: Francesc Carreras, Department of Applied Mathematics II and School of Industrial Engineering of Terrassa, Polytechnic University of Catalonia, ETSEIT, Colom 11, E-08222 Terrassa, Spain. Fax: +34-35-205-2549; E-mail: francesc.carrreras@upc.es

MANFRED J. HOLLER and STEFAN NAPEL

MONOTONICITY OF POWER AND POWER MEASURES

ABSTRACT. Monotonicity is commonly considered an essential requirement for power measures; violation of local monotonicity or related postulates supposedly disqualifies an index as a valid yardstick for measuring power. This paper questions if such claims are really warranted. In the light of features of real-world collective decision making such as coalition formation processes, ideological affinities, a priori unions, and strategic interaction, standard notions of monotonicity are too narrowly defined. A power measure should be able to indicate that power is non-monotonic in a given dimension of players' resources if – given a decision environment and plausible assumptions about behaviour – it *is* non-monotonic.

KEY WORDS: Power measures, monotonicity, voting

1. INTRODUCTION

Consider a set of players who jointly take decisions under a given set of rules. For example, the rules may specify that any player $i = 1,\ldots,n$ has a specific voting weight w_i and that a collective decision requires enough supporters such that their total weight exceeds a decision quota d. Power indices address the question of how much power collective decision rules, e.g. a weighted voting rule, award to individual players: Is player i more or less powerful than player j, and by how much?

Let p_i be the power value assigned to player i by a power index. A power distribution indicated by this index for a given rule involving voting weights is *locally monotonic* if $w_i > w_j$ implies $p_i \geq p_j$, i.e., a voter i who controls a larger share of vote does not have less power than a voter j with a smaller voting weight. An index which for any decision rule produces locally monotonic power distributions is said to satisfy *local monotonicity* or be a *locally monotonic index*. An index which violates local monotonicity exhibits a so-called *weighted voting paradox*, i.e. a player with greater weight having smaller power com-

Theory and Decision **56:** 93–111, 2004.

pared to another player, for at least some – though not necessarily many – weight assignments and quotas.

Monotonicity is commonly considered an essential requirement for power measures. Felsenthal and Machover (1998, p. 221ff), for instance, argue that any a priori measure of power that violates local monotonicity is "pathological" and should be disqualified as serving as a valid yardstick for measuring power. However, several notions of monotonicity have been defined and no consensus has been reached about how devastating a violation of a given type of monotonicity is in itself and in comparison to violations of other types. In our view, this has a good reason. Namely, the correct notion of monotonicity and whether monotonicity of a power distribution – and hence an index – is meaningful at all is highly context-dependent.

Typically, there is much more to a decision rule than weights and quota. Players face particular opportunities (and restrictions) of coordinating their support for – or opposition to – a given decision proposal. They can to different extents and by many different ways influence which possible decisions are considered in the first place. These aspects of the environment in which an at first sight very narrowly defined decision rule is applied – viewed from the different perspectives of individual players – constitute *resources* in a wider sense. Players can have very different ways to employ them, e.g. in a utility-maximizing way or using a particular adaptive heuristic. We will argue that a power measure should be able to indicate that power is non-monotonic in certain resource dimensions if – in view of plausible behavioural assumptions – it *is* non-monotonic in these resources (also see Holler and Napel, 2004).

The remainder of the paper is organized as follows: Section 2 discusses several postulates that are closely related to local monotonicity. Section 3 deals with special cases of voting bodies in which power is (designed to be) proportional to voting weights, guaranteeing monotonicity. In contrast, Section 4 considers several aspects of real-world collective decision situations and behavioural assumptions which can lead to violations of monotonicity. Section 5 discusses implications of a redistribution of votes and of new members joining a decision body, before Section 6 concludes.

2. MONOTONICITY POSTULATES

There are at least three postulates in addition to the requirement of local monotonicity which try to capture intuitive notions of monotonicity in the context of power: The *dominance postulate* requires $p_i \geq p_j$ whenever i dominates j, i.e., if it is true that for every coalition S such that j is not in S and the union of S and $\{j\}$ is a *winning coalition*, i.e. is able to take a decision, the union of S and $\{i\}$ is also a winning coalition.[1]

The *transfer postulate* demands that the power of any voter i given a weighted voting rule should not increase if i donates a part of his or her voting rights to another voter j provided that i is the sole donor. If this postulate is violated, then the corresponding power measure suffers from the *donation paradox* for some – though not necessarily many – weighted voting rules: i gains power by giving away votes to j.

Finally, the *bloc postulate* stipulates that the power of a merged entity $\{i, j\}$ is strictly larger than the power of player i before the merger provided that j is not a dummy, i.e. is a crucial member of at least some feasible coalition in the sense that he or she can turn it from winning to losing by leaving.

The definition of local monotonicity makes it particularly clear that the vote distribution is supposedly the measure rod with which a given power distribution must compare well. This is inherent also to the three mentioned principles. Thus, more or less explicitly, the monotonicity discussion is based on the premise that, in the case of weighted voting, there is a close relationship between power and votes, where votes represent resources. At the same time it is widely agreed that the vote distribution is a poor proxy for the distribution of voting power – this is, in fact, the main motivation behind the study of power indices.

It is well-known that the *normalized* (relative) Banzhaf index (Banzhaf, 1965) violates the transfer postulate and the bloc postulate, but satisfies the dominance postulate and local monotonicity. As a consequence, Felsenthal and Machover (1995, p. 225) conclude that the normalized Banzhaf index "must, at best, be regarded as seriously flawed". The non-normalized (absolute) Banzhaf index and the Shapley–Shubik index (1954) obey all four principles, while the Deegan–Packel (1979) index and the Public Good Index (Holler and Packel, 1983) violate all four principles.

If voting weights are identified with power values then the above postulates are trivially satisfied, and none of the paradoxes related to their violations is possible. For example, if we simply equate power with voting weights then, by definition, giving away voting weights reduces power and the donation paradox can never be observed.

If any violation of local monotonicity or one of the postulates is regarded as an indicator of a serious flaw in the definition of an index, one may ask: Why do we not simply take the voting weights (or their ratio to the decision quota) as power measures instead of more sophisticated measures that may suffer from various paradoxes? Of course, many different weight configurations can yield exactly the same sets of winning coalitions (and losing coalitions) and what matters to a coalition in the context of weighted voting is not by how much it is above or below the quota – just if it is. Taking any *arbitrary* element of the equivalence class of weights yielding a given set of winning coalitions is therefore clearly misleading and must produce inconsistencies. One could avoid this, for example, by using the minimal representation of a voting game (see Ostmann, 1987). This will always satisfy local monotonicity. Does the distance between values of an index, which always respects the weight ordering, really contribute to the understanding of power?

Perhaps violations of local monotonicity and the various paradoxes tell us a more substantial story about the properties of power in voting bodies? Brams and Affuso (1976, p. 52f) observe: "Given the widespread use and acceptance of power indices, we believe that an aberration they show up must be taken seriously. Instead of thinking of the paradox of new members as 'aberrant', however, we prefer to view it as an aspect of voting power whose existence would have been difficult to ascertain in the absence of precise quantitative concepts... It is a limitation in our thinking and models, not an aberration in the phenomenon, that has heretofore led us to equate power and size."

The *paradox of new members* they refer to occurs if the power value of an incumbent voter i increases when a new voter j enters the voting body and the votes of all incumbent voters and the percentage quota remain unaltered (implying, however, that the relative share of votes of the incumbent voter decrease).

Examples show that different power indices may not agree concerning the paradox of new members (see Brams and Affuso, 1976). For instance, for a specific voting game, the Shapley–Shubik index may indicate a paradox of new members while the (normalized) Banzhaf index does not. In principle, however, both measures can point out the existence of this paradox. This is, in our view, more an advantage than a flaw. A similar thing can be argued for violations of local monotonicity, which cannot happen for the Shapley–Shubik and Banzhaf indices but for the Deegan–Packel and Public Good indices.

Holler and Napel (2004) discuss in more detail whether local monotonicity is a property of the power distribution of a specific voting game, as represented by the chosen power measure, or whether it is a property of power per se. In the latter case, any reasonable power measure has to satisfy the local monotonicity axiom. If we accept the former perspective, then a measure which does not allow for a violation of local monotonicity is inappropriate to express this dimension of power. Only if the measure is able to indicate non-monotonicity it can serve to answer how the voting game has to be designed such that power values are monotonic in voting weights. Measures which produce values close to voting weights, irrespective of the vote distribution and the decision rule, and do not indicate any possible reversal of order (i.e. non-monotonicity) are rather useless instruments from this perspective.

3. STRICT PROPORTIONAL REPRESENTATION

Lionel Penrose conjectured already in the 1940s that in a weighted voting game with simple majority decision rule the ratio between the power values, measured by what later became known as the Banzhaf index, of any two voters converges to the ratio between their voting weights if the total number of voters goes to infinity. In a recent paper, Lindner and Machover (2004) provide sufficient conditions for this to be true. Strict proportionality between power and voting weights – or actually $p = w$ after rescaling – is the strongest form of monotonicity. Obviously, local monotonicity and the three monotonicity postulates or principles described above are immediately satisfied if strict proportionality holds.

In most real-world applications, however, we deal with a finite number of players which is often too small for these convergence results to matter. Then, the power distribution generically differs from the seat distribution. There is a substantial literature which discusses the problem of *equating* power with votes under the label of "fair allocation of votes" (see, e.g., Laruelle and Widgrén, 1998; Sutter, 2000; Laruelle, 2001; Leech, 2003). The starting point of this literature is that a (substantial) deviation of power from votes may be regarded as both an undesirable intransparency of the decision rule given the common misperception of power as proportional to weight, and – in a loose sense – unfair to players who get less out of their vote share than others. One can therefore consider redistributing votes such that $p = w$ is approximated "as closely as possible".

Even for large n, there are only finitely many power allocations p that can result from applying any of the established indices. In particular for small numbers of players the scope for achieving $p = w$ or, generally, for choosing voting weights $w = (w_1,\ldots,w_n)$ such that the vector $p = (p_1,\ldots,p_n)$ which expresses the power values assigned to the n agents of the voting game $v = (d; w)$ equals a design vector $k = (k_1,\ldots,k_n)$ which describes an exogenously given – perhaps particularly "fair" – allocation of voting power to the n agents, is very limited because of this discreteness.

Shapley (1962) proved that *strict proportionality* is obtained, i.e. $k = p = w$, if for any given w, quota d is uniformly distributed over the interval (0,1) and power is measured by the

Shapley–Shubik index. Dubey and Shapley (1979) show that this is also true when power is measured by the Banzhaf index. Holler (1985) and Berg and Holler (1986) apply the randomized decision rule principle to discrete probability distributions and small n, in order to achieve strict proportionality.

There are a few real-world applications of randomised collective decision making. For example, the recently proposed compromise concerning the composition of the Governing Council of the European Central Bank after expansion of the euro zone is that member countries get a seat on a rotating basis for a length of time corresponding to pre-specified *time shares*. From an a priori perspective these time shares can be interpreted as weights which correspond to the probability of having a vote on the issue arising at a random point of time. Decisions in the ECB Governing Council would according to the proposal be taken by simple majority among present Council members. This ensures that each member of the euro zone has power in strict proportion to its time weight.

However, to randomize not only in case of ties but as an essential part of a decision rule invites a number of objections – in particular from those unlucky players who happen to have no say on a given issue. If the principle of randomised decision rules were accepted in general, it would even be possible to randomly choose a dictator, with probabilities corresponding to the desired a priori power allocation – arguably the most straightforward way of guaranteeing monotonicity and even strict proportionality. This illustrates the high price that monotonicity can have. It might be one reason why we often find that strict proportionality is not satisfied in reality and why power indices are needed to illustrate corresponding distortions.

4. PERSPECTIVES ON LOCAL MONOTONICITY

Given the multitude of power measures that have been proposed in the literature, a possible strategy is to choose a favourite power measure and to try to convince others to share this choice. An alternative is to accept the multitude of measures and their interpretations and select an appropriate measure in concurrence with the intuition possibly based on the accompanying stories. A third alternative is to define discriminating properties, possibly in form of postulates or axioms, which a power measure has to satisfy in order to qualify as "appropriate". Local monotonicity has been proposed to be such a property.

Local monotonicity is an implication of *desirability* as proposed by Isbell (1958). This property formalizes that a voter i is at least as desirable as a voter j if for any coalition S, such that the union of S and $\{j\}$ is a winning coalition, the union of S and $\{i\}$ is also winning. Freixas and Gambarelli (1997) use desirability to define reasonable power measures. Since both the Deegan–Packel index and Public Good Index violate local monotonicity (see Holler, 1982; Holler and Packel, 1983), they also violate the desirability property. For example, given the vote distribution $w = (35, 20, 15, 15, 15)$ and a decision rule $d = 51$, the corresponding values of the Public Good Index are equal to: $h(d,w) = (16/60, 8/60, 12/60, 12/60, 12/60)$. A comparison of w and $h(d,w)$ shows a violation of local monotonicity. Because of the basic principles underlying the Public Good Index, which derive from the notion of a pure public good, (i.e. non-rivalry in consumption and exclusion of free-riding), only *decisive sets* (i.e. strict minimum winning coalitions) are considered when it comes to measuring power. All other coalitions are either losing or contain at least one member which does not contribute to winning. If coalitions of the second type are formed, then it is by luck, similarity of preferences, tradition, etc. – *but not because of power*.

It should be emphasized that the Public Good Index does not claim that only decisive sets will be formed but it suggests that that only decisive sets should be taken into consideration when it comes to measuring power and the outcome of collective decision making is a public good. As a consequence, each decisive set stands for a different kind of public good and thus alternative public goods can be characterized by the decisive sets which support them.

Some authors have explicitly specified coalition formation processes that can motivate non-monotonic indices. These processes naturally imply particular weights or probabilities for the ex post power enjoyed by a given player once a particular winning coalition has been formed. For example, Brams et al. (2003) consider simple majority voting with players who each have a linear preference ordering over possible coalition partners, and study two related coalition formation processes with significant empirical support. Specifically, they investigate a *fallback process* in which players seek coalition partners in descending order until a winning coalition of mutually "acceptable" players is established, and a *build-up process* which augments the fallback version by the requirement that no player outside the established coalition is strictly preferred by some coalition member to one of the insiders.

The probability that a coalition of size s is formed, assuming that all strict preference ordering are equally likely, turns out to be bimodal with peaks at the simple majority and unanimity. For given preference ordering, only particular coalitions of a given size s will be formed. It can, e.g., be the case that a comparatively small winning coalition is stable while a larger is not. Similar observations could be made for weighted majority voting. Power, under non-trivial coalition formation processes, therefore cannot be expected to be always monotonic.

Alonso-Meijide and Bowles (2003) make similar observations in their detailed analysis of voting power in the International Monetary Fund (IMF). They take an important institutional feature into account that has been neglected in other studies. The large number of 184 IMF member countries necessitates the a priori formation of 24 groups, each represented by a single director on the IMF's decision-making Executive Board. Member state's total power – resulting from power inside the group and the group's power in the Executive Board – can be derived using Owen's (1977, 1982) framework for power measurement with *a priori unions*. Alonso-Meijide and Bowles (2003) use sophisticated methods for the efficient computation of a priori union power indices. They evaluate three different measures of total power in the IMF – corresponding to alternative measures at the intra- and inter-group levels. The *Banzhaf–Owen index* applies the reasoning behind the Banzhaf index at both levels. It produces non-monotonic power indications, e.g. for Belgium and India when a European constituency that aggregates weights of European Union members and a quota of $d = 85\%$ are considered. Moreover, it has the additional "drawback" of failing to satisfy symmetry for a priori unions of equal weight, i.e. they may have unequal aggregate power.

The *Owen index*, which follows the weighting of marginal contributions underlying the Shapley–Shubik index, is symmetric but it fails to be monotonic nevertheless. This is also true for a new index introduced by Alonso-Meijide and Bowles, which distributes intra-group power according to the Shapley–Shubik index but inter-group power according to the Banzhaf index. The alternative *v-composition framework* for measuring total power given hierarchical decision levels, modelling decision-making in the groups as separate simple games and taking the Executive Board to be their composition, produces non-monotonic power distributions, too. In other words: Non-monotonicity becomes a very persistent feature of power as soon as the assumption of completely independent random yes–no decisions is replaced by features of real-world institutions.

If yes–no decisions concern particular proposals in a possibly multi-dimensional policy space, the probability of a given player being pivotal and thus having ex post power depends on his or her own position or most-preferred alternative in the policy space as well as those of the other players. Suppose that player i's most-preferred alternative or ideal point lies inside the convex hull of the ideal points of the other players. Then there exists no proposal that could make i the least or most enthusiastic player, in contrast to the players on the boundary of the convex hull. Considering random proposals, player i will therefore be the pivot player more often than players on the boundary who have the same weight. In fact, i will be in a powerful pivot position more often even than boundary players with greater weight, provided that the weight difference is not too big. This is formally captured by the *Owen-Shapley spatial power index* (see Owen, 1971; Shapley, 1977; Owen and Shapley, 1989), which modifies the Shapley–Shubik index in a way that takes account of ideological proximity among players. For transparent and very good reasons, this index violates local monotonicity.

Braham and Steffen (2002) demonstrate that applications of Straffin's (1977) *partial homogeneity approach*, which concerns particular assumptions about probabilistic yes–no decisions in Owen's (1972) *multilinear extension* of weighted voting games, do not always produce results consistent with local monotonicity. This is because partial homogeneity treats players asymmetrically in a special way and, as mentioned, the power of a voter *i* by definition depends not only upon the number of coalitions for which *i* is critical but also upon the probabilities by which the various coalitions arise. Braham and Steffen argue that Straffin's partial homogeneity approach is not less a priori than the Banzhaf index and the Shapley–Shubik index. The partial homogeneity approach can, in fact, be interpreted as a combination of the Banzhaf index and the Shapley–Shubik. These two indices satisfy local monotonicity, although their original axiomatization does not include local monotonicity.

Originally, the Deegan–Packel index and the Public Good Index derive from an axiomatic approach; probabilistic arguments do not necessarily apply to these measures. However, if we generalize Owen's multilinear extension framework to allow for the Public Good Index by applying zero probabilities to winning coalitions with surplus players, then the a priori argument which Braham and Steffen (2002) provide for the partial homogeneity measure applies also to the Public Good Index.

Returning to the above-mentioned "problem" of having a multitude of power measures to choose from – which can indeed assign players not only different index values for a given decision body, but also produce a different power ranking – it is noteworthy that the discriminating power of various monotonicity postulates is in any case questionable. Laruelle and Valenciano (2003) demonstrate that a *large class of power indices*, which measure the probability of players being decisive or pivotal given a weighted voting rule and probabilistic assumptions about their yes-or-no votes, passes the monotonicity "tests" that are commonly advocated in order to screen good power indices from bad ones. Moreover, an *even larger class of success indices*, measuring the probability of agreeing with the collective decision no matter if the considered player has contributed to it or not, passes exactly the same tests. In other words, postulates designed to filter out the "ideal index" apparently formalise monotonicity notions applying much more to success than to power.

5. REDISTRIBUTION OF VOTES AND NEW MEMBERS

When it comes to monotonicity of power with respect to voting weights, it is important to note that none of the existing measures guarantees that the power measure of player i will *not* decrease if his or her voting weight increases (also see Holler, 1998). Fischer and Schotter (1978) demonstrate this result (i.e., *the paradox of redistribution*) for the Shapley–Shubik index and the normalized Banzhaf index (see also Schotter, 1982). More specifically, take the voting game $v = (0.70, 0.55, 0.35, 0.10)$. The corresponding values for the normalized Banzhaf and Shapley–Shubik indices are $\beta(v) = \Phi(v) = (1/2, 1/2, 0)$. Now let's assume that the vote distribution $w = (0.55, 0.35, 0.10)$ changes to $w^{\circ} = (0.50, 0.25, 0.25)$ while the decision rule $d = 0.70$ remains unchanged. Then the resulting voting game $v^{\circ} = (0.70, 0.50, 0.25, 0.25)$ corresponds to the normalized Banzhaf index $\beta(v^{\circ}) = (3/5, 1/5, 1/5)$ and to the Shapley–Shubik index $\Phi(v^{\circ}) = (2/3, 1/3, 1/3)$. Both measures show that although the first voter's voting weight decreased from 0.55 to 0.50, his power increased, irrespective whether measured by Banzhaf or Shapley–Shubik.

Of course, there are voting games which are robust against the paradox of redistribution which we just demonstrated. For example, for the voting game $v' = (0.70, 0.55, 0.25, 0.20)$ there is no alternative vote distribution so that the paradox of redistribution prevails if power is measured by Banzhaf or Shapley–Shubik. This raises the question how important the paradox of redistribution phenomenon is. Fischer and Schotter

(1978) give some results which indicate that it has some substance for larger voting bodies. They prove the following propositions:

PROPOSITION 1. *For voting bodies with $n = 6$, a paradox of redistribution is always possible no matter what initial vote distribution exists, if power is measured by the Banzhaf index.*

PROPOSITION 2. *For voting bodies with $n = 7$, a paradox of redistribution is always possible no matter what initial vote distribution exists, if power is measured by the Shapley–Shubik index.*

PROPOSITION 3. *If $d = 1/2$, i.e., a simple majority decision rule applies and $n = 4$, then a paradox is always possible, irrespective of the initial distribution.*

Given the popularity and widespread dissemination of simple majority voting, the result in Proposition 3 should be rather alarming for those who are worried about non-monotonicity of all kinds. Of course, none of the above proposition says that the paradox of redistribution is a frequently observed phenomenon. However, a comparison of Proposition 1 and Proposition 2 suggests that the paradox is more frequent when power is measured by the Banzhaf than by Shapley–Shubik index. In other words, there is some evidence that the Banzhaf index is more liable to non-monotonicities than the Shapley–Shubik index. However, there can be no proof of this proposition – mainly because the argument becomes circular if power relations in the voting body are expressed by the corresponding power measure only: power is then what the index measures, and if the index indicates monotonicity then power is monotonic.

The paradox of redistribution stresses the fact that power is a social concept: if we discuss the power of an individual member of a group in isolation from his or her social context, i.e. related only to his or her individual resources, we may experience all sorts of paradoxical results. It seems that sociologists are quite aware of this problem and non-monotonicity of an individual's power with respect to his or her individual resources does not come as a surprise to them (see, e.g., Caplow, 1968).

Political scientists, however, often see the non-monotonicity of power as a threat to the principle of democracy. To them it is hard to accept that increasing the number of votes a group has could decrease its power, although it seems that there is ample empirical evidence for it (see Brams and Fishburn, 1995, for references). In general, economists also assume that controlling more resources is more likely to mean more power than less. However, they also deal with concepts like monopoly power, bargaining, and exploitation which stress the social context of power and the social value of resources (assets, money, property, etc.).

The *paradox of redistribution* is closely related to the *paradox of new member*. Felsenthal and Machover (1995) give an example in which the decision rule d is the same for the game before and after entry of a fifth voter j. They adjust the vote ratios so that the vote shares add up to one before and after entry of j. Thus, as a consequence the shares of the incumbent voters have to decrease. The games before entry, v', and after entry, v'', are

$$v' = (0.51, 0.30, 0.30, 0.30, 0.10) \quad \text{and}$$

$$v'' = (0.51, 0.15, 0.15, 0.15, 0.05, 0.50).$$

The *paradox of new members* is obvious. In the game v', having 10 percents of the votes, player 4 is a dummy. In the game v'', now having only 5 percents of the votes, player 4 can form a coalition with the entrant player 5 who controls 50 percent of the votes. Felsenthal and Machover (1995, p. 222) argue that "any reasonable index of relative voting power" has to display the paradox of new members in this case. However, note that v' is equivalent to $v°$ which assigns a vote share of 0.00 to player 5 so that $v° = (0.51, 0.30, 0.30, 0.30, 0.10, 0.00)$ alternatively describes the game before entry of player 5. Now if we compare $v°$ to v'' we clearly see that the *paradox of redistribution* prevails. Player 4 loses half of his vote share but is no longer a dummy in game v''.

Because of the adjustment of seat shares and by keeping the decision rule unchanged, the example of Felsenthal and Machover captures both the *paradox of redistribution* and the *paradox of new member*. Brams and Affuso (1976) originally discussed the *paradox of new member* for the addition of one or two players to the votes of the incumbents.

Brams and Affuso (1976, p. 52) observe that the probabilities for occurrence of the paradox of new members "are high in relatively small weighted voting bodies". They argue that whether "there exists a voting body invulnerable to the paradox is of less practical import than the probability of occurrence of the paradox" (p. 50). The paradox of new members shows that the relative number of swings of a weighted voter i with a constant number of seats can increase if new weighted voters enter the voting body, and thus the relative vote share of i decreases. If power expresses the potential to form and to contribute to coalitions and thereby to control the outcome, then this paradox does not come as a surprise and an index which is not equipped to indicate the paradox seems inadequate to discuss the properties of the power relations in this case. In the end, an adequate power measure should clarify the properties of the game so that, for example, a disfavoured player (or unhappy designer) can change the game.

Power indices with a strong sensitivity to monotonicity can also be of help for a more abstract analysis of decision situations with respect to power. Myerson (1999, p. 1080) argues that "the task for economic theorists in the generations after Nash has been to identify the game models that yield the most useful insights into economic problems. The ultimate goal of this work will be to build a canon of some dozens of game models, such that a student who has worked through the analysis of these canonical examples should be prepared to understand the subtleties of competitive forces in the widest variety of real social situations." What can be said of understanding the subtleties of competitive forces also applies to power. The analysis of alternative social situations by means of power measures complements more direct approaches to enhance the understanding of power, its sources and its consequences.

6. CONCLUDING REMARKS

The fundamental problem that leads to non-monotonicity of power distributions for a given index and special decision rules should not be attributed per se to a supposed inappropriateness of the index. Rather, the inappropriateness lies in the description of a social or economic situation involving rational or boundedly rational agents who have to reach a collective decision by merely a vector of weights *w* and a quota *d*. All established indices are based on additional assumptions about the situation and agents either explicitly or, more often, implicitly (e.g., in the form of axioms about the behaviour of the index or probabilistic assumptions about the behaviour of players).

The high degree of abstraction entailed by the Spartan weight-quota framework therefore almost implies that indices will yield power distributions that can be considered "paradoxical" from some point of view which is inconsistent with these assumptions, at least for situations in which the inconsistency is so pronounced that it is in fact desirable to observe the supposed "paradox".

What is "paradoxical" under the implicit assumption of independent random votes on an exogenously given random proposal can be perfectly consistent with or directly called for by common sense in the context of a given coalition formation process, affinities between agents implied e.g. by certain positions in a policy space, or the necessity to form a priori unions – and vice versa! A claim that an index, such as the Public Good Index or the Owen-Shapley spatial power index, should be discarded because it violates local monotonicity or some monotonicity postulate amounts to a claim that the set of assumptions about the decision situation and players' behaviour which are underlying it are invalid. In our opinion, such strong assertions are unjustified in general. They may be substantiated in the analysis of *particular* real-world decision environments, which may be inconsistent with the assumptions underlying a given index. Ironically, it is the puristic weight-quota framework with stochastically independent yes-no decisions on unspecified proposals, for which local monotonicity is indeed an understandable concern, which has the biggest problems in finding real-world situation that 'fit' and in convincing decision makers of its relevance inside real-world institutions to which power indices are most commonly applied.

NOTES

1. A different notion of dominance entails player *j* contributing to a coalition in the sense of turning it from losing into winning *only in the presence of i* while *i* contributes also to coalitions that do not involve *j* (see Napel and Widgrén, 2001).
2. The probabilistic approach to power measurement has been generalized by Napel and Widgrén (2004) to a much wider class than commonly considered. They propose to calculate a priori power as expected a posteriori power, which in turn is inferred from the collective decision's sensitivity to action or preference trembles by individual players. This includes traditional indices as special cases, but can assess power derived by strategic behaviour in non-trivial decision-making procedures, too. Interaction between the general *decision behavior* of players and the *decision situation* described by, among other things, voting weights can explicitly be accounted for. Alternative decision situations and assumptions on expected behavior can imply monotonicity or non-monotonicity of power in voting weights.

REFERENCES

Alonso-Meijide, J.M. and Bowles, C. (2003), Power indices restricted by a priori unions can be easily computed and are useful: a generating function-based application to the IMF. Discussion Paper.

Banzhaf, J.F. (1965), Weighted voting doesn't work: a mathematical analysis, *Rutgers Law Review* 19, 317–343.

Braham, M. and Steffen, F. (2002), Local monotonicity and Straffin's partial homogeneity approach to the measurement of voting power. Discussion Paper of the Institute of SocioEconomics. University of Hamburg, Hamburg.

Berg, S. and Holler, M.J. (1986), Randomized decision rules in voting games: A model for strict proportional power, *Quality and Quantity* 20, 419–429.

Brams, S.J., Jones, M.A. and Kilgour, D.M. (2003), Forming stable coalitions: the process matters. Discussion Paper.

Brams, S.J. and Affuso, P.J. (1976), Power and size: a new paradox, *Theory and Decision* 7, 29–56.

Brams, S.J. and Fishburn, P.C. (1995), When size is liability? Bargaining power in minimal winning coalitions, *Journal of Theoretical Politics* 7, 301–316.

Caplow, T. (1968), Two against one: coalitions in triads. Englewood Cliffs, N.J.: Prentice-Hall.

Deegan, J. Jr. and Packel, E.W. (1979), A new index of power for simple n-person games, *International Journal of Game Theory* 7, 113–123.

Dubey, P. and Shapley, L.S. (1979), Mathematical properties of the Banzhaf index, *Mathematics of Operations Research* 4, 99–131.

Felsenthal, D. and Machover, M. (1995), Postulates and paradoxes of relative voting power – A critical review, *Theory and Decision* 38, 195–229.

Felsenthal, D. and Machover, M. (1998), The measurement of voting power. Theory and practice, problems and paradoxes (Edward Elgar, Cheltenham).

Fischer, D. and Schotter, A. (1978), The inevitability of the paradox of redistribution in the allocation of voting weights, *Public Choice* 33, 49–67.

Freixas, J. and Gambarelli, G. (1997), Common internal properties among power indices, *Control and Cybernetics* 26, 591–603.

Holler, M.J. (1982), Forming coalitions and measuring voting power, *Political Studies* 30, 262–271.

Holler, M.J. (1985), Strict proportional power in voting bodies, *Theory and Decision* 19, 249–258.

Holler, M.J. (1998), Two stories, one power index, *Journal of Theoretical Politics* 10, 179–190.

Holler, M.J. and Napel, S. (2004), Local monotonicity of power: axiom or just a property? *Quality and Quantity* (forthcoming).

Holler, M.J. and Packel, E.W. (1983), Power, luck and the right index, Zeitschrift für Nationalökonomie, *Journal of Economics* 43, 21–29.

Isbell, J.R. (1958), A class of simple games, *Duke Mathematics Journal* 25, 423–439.

Laruelle, A. (2001), Implementing democracy in indirect voting processes: the Knesset case, in M.J. Holler and G. Owen (eds), *Power Indices and Coalition Formation*, London: Kluwer, Boston, Dordrecht.

Laruelle, A. and Valenciano, F. (2003), Some voting power postulates and paradoxes revisited. Discussion Paper.

Laruelle, A. and Widgrén, M. (1998), Is the allocation of voting power among the EU states fair?, *Public Choice* 94, 317–339.

Leech, D. (2003), Power indices as an aid to institutional design: the generalised apportionment problem, in M.J. Holler, H. Kliemt, D. Schmidtchen and M.E. Streit (eds), *European Governance* (Jahrbuch für Neue Politische Ökonomie 22) (Mohr Siebeck, Tübingen).

Lindner, I. and Machover, M. (2004), L.S. Penrose's limit theorem: proof of some special cases, *Mathematical Social Sciences* 47, 37–49.

Manin, B. (1997), *The principles of representative government*, Cambridge: Cambridge University Press.

Myerson, R.B. (1999), Nash equilibrium and the history of economic theory, *Journal of Economic Literature* 37, 1067–1082.

Napel, S. and Widgrén, M. (2001), Inferior players in simple games, *International Journal of Game Theory* 30, 209–220.

Napel, S. and Widgrén, M. (2004), Power measurement as sensitivity analysis – A unified approach, *Journal of Theoretical Politics* (forthcoming).

Ostmann, A. (1987), On the minimal representation of homogeneous games, *International Journal of Game Theory* 16, 69–81.

Owen, G. (1971), *Political games.* Naval Research Logistics Quarterly, Vol 18, pp. 345–355.

Owen, G. (1972), Multilinear extensions of games, *Management Science* 18, 64–79.

Owen, G. (1977), Values of games with a priori unions, in R. Henn and O. Moeschlin (eds), *Mathematical Economics and Game Theory*, Berlin: Springer.

Owen, G. (1982), Modification of the Banzhaf-Coleman index for games with a priori unions, in M.J. Holler (ed), *Power, Voting and Voting Power*, Würzburg and Wien: Physica-Verlag.

Owen, G. and Shapley, L.S. (1989), Optimal location of candidates in ideological space, *International Journal of Game Theory* 18, 339–356.

Schotter, A. (1982), The paradox of redistribution, in M.J. Holler (ed), *Power, Voting and Voting Power*, Würzburg and Vienna: Physica-Verlag.

Shapley, L.S. (1962), Simple games: An outline of the descriptive theory, *Behavioral Science* 7, 59–66.

Shapley, L.S. (1977), *A comparison of power indices and a non-symmetric generalization*, Paper P-5872, Santa Monica, CA: Rand Corporation.

Shapley, L.S. and Shubik, M. (1954), A method of evaluating the distribution of power in a committee system, *American Political Science Review* 48, 787–792.

Straffin, P.D. (1977), Homogeneity, independence, and power indices, *Public Choice* 30, 107–118.

Sutter, M. (2000), Fair allocation and re-weighting of votes and voting power in the EU before and after the next enlargement, *Journal of Theoretical Politics* 12, 433–449.

Addresses for correspondence: Manfred J. Holler, Institute of SocioEconomics (IAW), University of Hamburg, Von-Melle-Park 5, D-20146 Hamburg, Germany (Fax: +49-40-42838-3957; E-mail: holler@econ.uni-hamburg.de)

Stefan Napel, Institute of SocioEconomics (IAW), University of Hamburg, Von-Melle-Park 5, D-20146 Hamburg, Germany. E-mail: napel@econ.uni-hamburg.de

A. LARUELLE and F. VALENCIANO

ON THE MEANING OF OWEN–BANZHAF COALITIONAL VALUE IN VOTING SITUATIONS

ABSTRACT. In this paper we discuss the meaning of Owen's coalitional extension of the Banzhaf index in the context of voting situations. It is discussed the possibility of accommodating this index within the following model: in order to evaluate the likelihood of a voter to be crucial in making a decision by means of a voting rule a second input (apart from the rule itself) is necessary: an estimate of the probability of different vote configurations. It is shown how Owen's coalitional extension can be seen as three different normative variations of this model.

KEY WORDS: Owen's coalitional values, power indices, Banzhaf index

1. INTRODUCTION

Since Aumann and Drèze (1975) a variety of "coalitional values" have been proposed. A "coalitional value" is an evaluation for the different players of the prospect of engaging in a game situation described by a coalitional or transferable utility game and a "coalitional structure", this meaning a partition of the set of players into disjoint groups. Among these coalitional values two have been object of special attention and widely applied to real world situations: Owen's coalitional extensions to this class of situations of the Shapley (1953) value (Owen, 1977) and that of the Banzhaf (1965) index (Owen, 1982). In this note we concentrate on the second extension, to which for the sake of brevity we will refer to here as the "Owen–Banzhaf coalitional value" (or "index" when restricted to voting rules, often described as simple games).[1] More specifically, we concentrate on its meaning in the context of voting situations, and discuss the possibility of accommodating this index within the following probabilistic model: in order to evaluate the likelihood of a voter to be crucial in making a decision by means of a

Theory and Decision **56:** 113–123, 2004.

voting rule a second input (apart from the rule itself) is necessary: an estimate of the probability of different vote configurations. It is shown how Owen's coalitional extension of the Banzhaf index can be seen as three different normative variations of this model.

2. BASIC BACKGROUND: VOTING RULES AND DECISIVENESS

We consider collective decision-making situations in which a set of voters makes decisions by means of a voting rule.[2] The voting rule specifies which vote configurations ("coalitions" is the usual term in cooperative games) are winning, that is, will entail the acceptance of the proposal, which otherwise would be rejected. If $N = \{1, \ldots, n\}$ denotes the set of *seats* (as well as the voters occupying them), and any vote different from "yes" is assimilated to "no", there are 2^n possible *vote configurations.* Each vote configuration can be represented by the set $S \subseteq N$ of "yes" voters. Thus an N-voting rule is fully specified by the set of winning configurations. It is usually assumed that: (i) $N \in W$; (ii) $\emptyset \notin W$ and (iii) If $S \in W$, then $T \in W$ for any T containing S; and (iv) If $S \in W$ then $N \backslash S \notin W$. The last condition, which prevents the possibility of a proposal and its opposite both being accepted if they are supported by two disjoint groups of voters, can be dropped in some cases. We will drop i's brackets in $S \backslash \{i\}$ and $S \cup \{i\}$.

As is well known a voting rule can also be represented by a simple coalitional game. An *N-coalitional game* is a pair (N,v), where $N = \{1, \ldots, n\}$ is a set of "players" and v is a map that assigns to each subset or coalition $S \subseteq N$ a real number or "worth" $v(S)$, such that $v(\emptyset) = 0$. A coalitional game v is *simple* if $v(S) = 0$ or 1. Then a voting rule W can also be represented by the simple coalitional game $v_W : 2^N \rightarrow R$, such that $v_W(S) = 1$ if $S \in W$ and $v_W(S) = 0$ if $S \notin W$.

In order to assess the likelihood of a voter playing a relevant role in making a decision, the other voters' behaviour is determinant, but in general it is not known in advance how voters are going to vote. Here we assume that an estimate of the likelihood of the different vote configurations is available, that is, a probability $p(S)$ for each vote configuration S, so that $0 \leq p(S) \leq 1$ for any $S \subseteq N$, and $\sum_{S \subseteq N} p(S) = 1$. Then voter i's *decisiveness*, or probability of being decisive for making a decision by means of W, is given by

$$\Phi_i(W,p) := \text{Prob } (i \text{ is decisive}) = \sum_{\substack{S: i \in S \in W \\ S \setminus i \notin W}} p(S) + \sum_{\substack{S: i \notin S \notin W \\ S \cup i \in W}} p(S). \quad (1)$$

3. THE BANZHAF INDEX AND THE OWEN–BANZHAF COALITIONAL INDEX

Banzhaf (1965) proposes as an index of the capacity of a voter to be influential in making decisions by means of a given voting rule the number of "swings" or winning configurations in which such voter is crucial. That is, for a voter i in an N-voting rule W, it is given given by

$$\beta_i(W) := \#\{S \subseteq N : i \in S \in W \text{ and } S \setminus i \notin W\},$$

where # stands for the cardinal of the set. A "normalization" (see Owen, 1975; Dubey and Shapley, 1979) of this "raw" Banzhaf index is given by

$$Bz_i(W) = \frac{\beta_i(W)}{2^{n-1}}.$$

This meaningful "normalization" of the original Banzhaf index (which in what follows we will refer to as the "Banzhaf index") fits naturally into the model introduced in the previous section. In the absence of any information concerning the voters' behaviour, or in case of deliberately ignoring it if available with normative purposes, it seems natural to assume all vote configurations equally probable.[3] That is, taking

$$p^*(S) = \frac{1}{2^n}, \quad \text{for all } S \subseteq N.$$

Then we have:

$$\Phi_i(W,p^*) = \sum_{\substack{S:i\in S\in W\\ S\setminus i\notin W}} (p^*(S) + p^*(S\setminus i))$$

$$= \sum_{\substack{S:i\in S\in W\\ S\setminus i\notin W}} \frac{1}{2^{n-1}} = Bz_i(W). \tag{2}$$

In words: $Bz_i(W)$ gives the probability of voter i being decisive in a decision made by means of W if all vote configurations are equally probable. Note that (2) can be alternatively expressed in terms of the simple game v_W representing the rule W as

$$Bz_i(v_W) = \sum_{S:i\in S} \frac{1}{2^{n-1}}(v_W(S) - v_W(S\setminus i)), \tag{3}$$

and observe that, as pointed out by Owen (1975), (3) makes sense if v_W is replaced by an arbitrary coalitional game v. In this case $Bz_i(v)$, to which we will refer as the "Banzhaf semivalue" of player i in game v, gives the average marginal contribution of player i to the coalitions containing her.

Owen (1982) proposes an extension of this index to the case in which a "coalitional structure", interpreted as a form of *a priori* union of some subgroups of players, is given exogenously along with the voting rule.[4] An *N-coalitional structure* is a partition $\mathcal{B} = \{B_j\}_{j\in M}$, where $M = \{1,\ldots,m\}$, of the set of players N into m disjoint groups. For any $Q \subset M$, denote

$$N(Q) := \bigcup_{q\in Q} B_q. \tag{4}$$

Owen's "modification of the Banzhaf–Coleman index" according to his own terms, to which we refer here as the "Owen–Banzhaf coalitional index", for a voter i in bloc B_j is given by

$$OwBz_i[W;\mathcal{B}] := \frac{1}{2^{m+b_j-2}} \sum_{Q\subset M\setminus j} \sum_{\substack{K:\\ i\in K\subset B_j}} [v_W(N(Q)\cup K)$$

$$-v_W(N(Q)\cup(K\setminus i))], \tag{5}$$

where b_j denotes the cardinal of B_j. Observe that formula (5) makes also sense if v_W is replaced by an arbitrary coalitional game v.

In Owen (1982) two equivalent formulations of this index are given. Formula (5) can be directly interpreted as the ratio of vote configurations in which i is decisive (for the rule W) and in which no bloc different from B_j is broken, w.r.t. the total number of such vote configurations. A more complex formulation is obtained by applying twice the Banzhaf index or semivalue. Namely applying the Banzhaf semivalue to a B_j-game in which the worth of every coalition is given by the Banzhaf index of this coalition playing as a voter in a different M-voting rule. We will come back with more detail to this formulation in the next section.

4. INTERPRETATION OF THE OWEN–BANZHAF COALITIONAL INDEX

In the context of voting, ex ante unions arise naturally (parties, blocs, etc.) and may constrain the actual vote configurations. If in addition to the voting rule a coalitional structure, interpreted as a form of ex ante union into subgroups of players, is given, the natural treatment (within the model introduced in Section 2) consists of restricting the class of probability distributions to those that assign probability zero to those configurations that "break" any coalition in this structure. However, if all voters vote in blocs, the situation can be interpreted as a *quotient M*-voting rule by means of which the blocs as single voters make decisions, whose winning configurations, using (4), are

$$W_B^M := \{Q \subset M \text{ s.t. } N(Q) \in W\}.$$

In order to assess the blocs' respective decisiveness, from a normative point of view corresponding to ignorance of any further information beyond the rule itself *and* the coalitional structure, we can assume that all vote configurations are equally probable. Thus, for each $j \in M$, $Bz_j(W_B^M)$ gives the probability of bloc j being decisive in the inter-blocs quotient rule. But the evaluation given by (5), based on the same two inputs, W and $\mathcal{B}$, is completely different. The aim of this paper is precisely to discuss its meaning.

As we show in this section, the Owen–Banzhaf coalitional index can be interpreted in three different ways as an answer to the following question: which is the relevance or decisiveness of each voter i if decisions are to be made by means of rule W and voters in every bloc in $\mathcal{B}$ other than the one containing i vote as a bloc (i.e., the vote does not split within any of the other blocs)? In other words: Which is the decisiveness imputable to every voter in the specified bloc under these conditions?

4.1. *A "Modified Banzhaf Index" of the Voting Rule*

The Owen–Banzhaf coalitional index, given by formula (5), admits the following interpretation consistent with the above described context: Fix a bloc B_j and assume that all blocs but B_j act as blocs, and every of these blocs votes "yes" with probability 1/2 and "no" with probability 1/2, while within B_j all vote configurations are equally probable (equivalently: every voter in B_j, independently, votes "yes" with probability 1/2). Thus we have:

PROPOSITION 1. *For any N-voting rule W and any N-coalitional structure* $\mathcal{B}$,

$$OwBz_i[W;\mathcal{B}] = \Phi_i(W, p_j^{\mathcal{B}}), \quad \text{for all } i \in B_j \in \mathcal{B}, \tag{6}$$

where $p_j^{\mathcal{B}}$ *denotes the distribution that assigns the same probability to all N-vote configurations that do not break any* $B_k \neq B_j$, *and zero to those which break any* $B_k \neq B_j$.

Thus, Owen–Banzhaf coalitional index appears as a "modified Banzhaf index" (i.e., replacing p^* by $p_j^{\mathcal{B}}$) of the original voting rule. The normative point of view is kept, but conditional to no bloc $B_k \neq B_j$ splitting the vote, that is,

$$p_j^{\mathcal{B}}(S) = Prob_{p^*}(S| \text{ blocs other than } B_j \text{ do not split}).$$

But note that $\Phi_i(W, p_j^{\mathcal{B}})$ fits general formulation (1) only partially because the probability distribution $p_j^{\mathcal{B}}$ depends on which bloc voter i belongs to. Thus, the evaluation of the decisiveness by the Owen–Banzhaf coalitional index *is based on different probability distributions for players in different blocs.*

4.2. *The Banzhaf Index of a Modification of the Voting Rule*

Another interpretation of formula (5) is obtained by treating blocs other than B_j as single voters, and voters in B_j as independent voters. Thus for any voter in B_j, the situation can be described by the $(M\backslash j) \cup B_j$ – voting rule

$$W_{\mathcal{B}}^{(M\backslash j)\cup B_j} := \{Q \cup K \text{ s.t. } Q \subset M\backslash j,\ K \subset B_j \text{ and } N(Q) \cup K \in W\}.$$

Then the Owen–Banzhaf index of any voter in B_j is given by the Banzhaf index of this voting rule.

PROPOSITION 2. *For any N-voting rule W and any N-coalitional structure* $\mathcal{B}$,

$$OwBz_i[W; \mathcal{B}] = Bz_i(W_{\mathcal{B}}^{(M\backslash j)\cup B_j}), \quad \textit{for all } i \in B_j \in \mathcal{B}. \quad (7)$$

In this case the probability distribution is always the one associated with the Banzhaf index, but *the voting rule considered for every bloc is different.*

4.3. *The (Extended) Banzhaf Index of an (Extended Notion of) Voting Rule*

The following interpretation is based on the alternative formulation proposed by Owen alluded formerly, which we examine now in more detail. Owen's second formulation of his coalitional index is obtained for each voter in each bloc by applying the Banzhaf semivalue to the coalitional game among the voters within that bloc in which the worth of each coalition is given by the Banzhaf index of this coalition when it plays the inter-blocs game with the other blocs replacing the bloc that contains it. That is, for each bloc B_j, and each subset $K \subset B_j$, consider the M-voting rule

$$W_{\mathcal{B},j,K}^{M} := \{Q \subset M \text{ s.t. } N(Q\backslash j) \cup K \in W\},$$

which is the voting rule that results from W if all blocs other than B_j act as blocs, and bloc B_j is replaced by coalition K acting as a bloc. Then assign to each K the Banzhaf index of voter j in the corresponding voting rule, $Bz_j(W^M_{B,j,K})$. In this way a new B_j-player coalitional game (not simple in general) is obtained, namely

$$v^{B_j}(K) := Bz_j(W^M_{B,j,K}), \quad \text{for each } K \subset B_j.$$

Finally the Banzhaf semivalue of player $i \in B_j$ in this new game yields the Owen-Banzhaf coalitional index, that is, for all $i \in B_j \in \mathcal{B}$

$$OwBz_i[W; \mathcal{B}] = Bz_i(v^{B_j}).$$

If one examines in detail this two-step formulation some difficulties of interpretation arise. In the first step the worth of every coalition in coalitional game v^{B_j} is given by the decisiveness (evaluated by the Banzhaf index) of that coalition acting as a bloc in voting rule $W^M_{B,j,K}$. Thus the Banzhaf *semivalue* of the resulting coalitional game is (see Section 3) the average marginal contribution of voter i to the decisiveness (in that way evaluated) of the coalitions in B_j containing i.

This construction takes us beyond the domain of voting rules (in general v^{B_j} is not a simple game) so that the second application of the Banzhaf *semivalue* may be questionable. In support of this application we show that v^{B_j} can be interpreted as a voting rule in an extended sense. Under this interpretation, the Banzhaf index (semivalue, strictly speaking) is applied *only once*.

First note that for any voting rule W, $v_W(S)$ can be interpreted as the increase in the probability of a proposal being accepted when passing from the sure configuration $\emptyset$ to the sure configuration S. In an ordinary voting rule this increase can only be 1 or 0, depending on whether S is winning or not. This idea can be applied when there exists a coalitional structure and it is assumed that voters in every other bloc than a fixed one never split the vote. Under this assumption we can assess "how much winning" is any coalition K within a bloc B_j ($K \subset B_j$) given that decisions by means of W involve voters outside B_j. This means that a same coalition in B_j may be "winning" or not depending on the vote configuration outside B_j. So we assign to each coalition K the increase in terms of probability that a proposal is accepted when coalition K votes "yes" (and $B_j \backslash K$

votes "no") compared with the probability when the whole bloc B_j votes "no". In the same normative spirit as before, for this assessment we assume all other blocs voting independently "yes" with probability 1/2. Under this assumption the increment of the probability of accepting a proposal when the behaviour *within* B_j switches from sure unanimous "no" to the sure configuration $K \subset B_j$ will *not* in general be neither 0 nor 1, as is the case in any ordinary voting rule, but would be given by

$$\sum_{\substack{Q \subset M \setminus j \\ K \cup N(Q) \in W}} \frac{1}{2^{m-1}} - \sum_{\substack{Q \subset M \setminus j \\ N(Q) \in W}} \frac{1}{2^{m-1}},$$

which is exactly $v^{B_j}(K)$. Thus the game v^{B_j} *can be interpreted as a voting rule in an extended sense.* Therefore it makes sense to take (as an *extended Banzhaf index*) the Banzhaf semivalue of this "voting rule" as a measure of the decisiveness of every voter in B_j in this context.

5. CONCLUSIONS

Thus we have provided three alternative interpretations of the Owen–Banzhaf coalitional index. In (6) the Owen–Banzhaf index appears as a "modified Banzhaf index" of the original rule, while in (7) as the Banzhaf index of a modified voting rule. Finally, in the last interpretation it appears as the Banzhaf semivalue (interpreted as an extended "index") of a coalitional game that can be interpreted as a voting rule in an extended sense.

In the three interpretations the Banzhaf index or semivalue (modified in the first case) is applied only once. But observe that in the three cases comparisons among the Owen–Banzhaf index of different voters make full and clear sense only within a same bloc. In formulation (6) p_j^B is *a probability distribution* over vote configurations in *W different for every bloc*. In (7) $W_B^{(M \setminus j) \cup B_j}$ represents a *different voting rule for every bloc*. In the last interpretation, v^{B_j} is a *different* (*extended*) *voting rule for every bloc*. In other words, the picture provided by the N-vector

$(OwBz_i[W; \mathcal{B}])_{i \in N}$ is in fact composed of m different pictures, every one of them telling something about a particular bloc. This should be taken into account in the applications of this index. Also note that in the three interpretations the point of view underlying the choice of the distribution of probability is normative. That is, all the vote configurations which are considered possible are assumed equally probable.

Owen's coalitional extension of the Banzhaf index provides an interesting example of an "*a priori*" assessment in which some information beyond the rule itself is taken into account: the coalitional structure. But in a very particular way: it is taken as part of the environment of a voter in a bloc (all the others will act as blocs) to assess the *a priori* decisiveness of every voter within her bloc given that context.

The reader may check that a similar interpretation for other coalitional values in the context of voting is problematic. In other words, the arguments given here do not adapt convincingly to other coalitional values in the context of voting situations.

ACKNOWLEDGEMENTS

This research has been supported by the DGES of the Spanish Ministerio de Ciencia y Tecnología under project BEC2000-0875, and by the Universidad del País Vasco under project 1/UPV00031.321-H-14872/2002. The first author also acknowledges financial support from the Spanish Ministerio de Ciencia y Tecnología under the Ramón y Cajal Program. Part of this paper was written while the second author was visiting the Departamento de Fundamentos del Análisis Económico, at the Universidad de Alicante, whose hospitality is gratefully acknowledged. The authors thank an anonymous referee for her or his comments.

NOTES

1. Owen (1982) refers to his extension as a "modification of the Banzhaf–Coleman index".

2. The basic framework, notation and terminology presented in this section are based on Laruelle and Valenciano (forthcoming).
3. As is well known, this is equivalent to assuming that every voter idependently votes "yes" with probability 1/2 and "no" with probability 1/2.
4. In fact, Owen (1982) extends it in the more general setting of coalitional games, but here we are interested in its meaning when applied to voting rules. Consequently we formulate it directly in terms of voting rules, though using the simple games notation in order to make the adaptation clear to the reader more familiar with this notation.

REFERENCES

Aumann, R.J. and Drèze, J. (1975), Solutions of cooperative games with coalitional structure, *International Journal of Game Theory* 4, 167–176.

Banzhaf, J. (1965), Weighted voting doesn't work: a mathematical analysis, *Rutgers Law Review* 19, 317–343.

Dubey, P. and Shapley, L.S. (1979), Mathematical properties of the Banzhaf power index, *Mathematics of Operations Research* 4, 99–131.

Laruelle, A. and Valenciano, F. (forthcoming), Assessing success and decisiveness in voting situations, *Social Choice and Welfare*.

Owen, G. (1975), Multilinear extensions and the Banzhaf value, *Naval Research Logistics Quarterly* 22, 741–750.

Owen, G. (1977), Values of games with a priori unions, in R. Hein and O. Moeschlin (eds), *Essays in Mathematical Economics and Game Theory* New York: Springer-Verlag, 76–88.

Owen, G. (1982), Modification of the Banzhaf–Coleman index for games with a priori unions, in M.J. Holler (ed.), *Power, Voting, and Voting Power*. Würzburg: Physica-Verlag, 232–238.

Shapley, L.S. (1953), A value for *n*-person games, *Annals of Mathematical Studies* 28, 307–317.

Address for correspondence: Annick Laruelle: Departamento de Fundamentos del Análisis Económico, Universidad de Alicante, Spain (E-mail: laruelle@merlin.fae.ua.es).

Federico Valenciano: Departamento de Economía Aplicada IV, Universidad del País Vasco, Bilbao, Spain (E-mail: elpvallf@bs.ehu.es).

MARCIN MALAWSKI

"COUNTING" POWER INDICES FOR GAMES WITH A PRIORI UNIONS

ABSTRACT. The classical Owen construction of the Shapley value for games with a priori unions is adapted to extend the class of "counting" power indices – i.e., those computed by counting appropriately weighted contributions of players to winning coalitions – to simple games with a priori unions. This class contains most well-known indices, including Banzhaf, Johnston, Holler and Deegan–Packel indices. The Shapley–Shubik index for simple games with a priori unions coincides with the (restriction of) Shapley value for such games obtained by Owen's method, but for all other indices obtained by normalization of probabilistic values our construction leads to indices different from those determined by Owen values.

KEY WORDS: power index, counting index, a priori unions

1. INTRODUCTION AND PREREQUISITES

Since their introduction in 1977 in the classical paper by Guillermo Owen, the notions of a game with a priori unions structure and of the Shapley value for such games have gained widespread recognition among game theorists. Also political scientists measuring "power" of participants in various decision-making assemblies became interested in the new concepts. However, as the Shapley value is one of many possible power measures (and not unanimously regarded the best one), there seems to be some demand from the side of political theory for other power indices defined on games with a priori unions. Owen's construction, however, cannot be readily applied here because it uses some auxiliary games which usually are not simple games even when the original game is simple.

In this paper we propose a partial solution – a method of extending every *counting* power index to simple games with a priori unions. The method is based on that of Owen (1977) but it is modified to work in the specific environment of simple games

Theory and Decision **56:** 125–140, 2004.

and (counting) power indices. Having introduced basic notions and notation in this section, we define the class of counting power indices in Section 2 and the proposed extension to games with precoalitions in Section 3, together with some examples. In the last section we compare indices for games with precoalitions obtained by this method with those resulting from straightforward generalization of Owen's construction in the case when the underlying index is a restriction of a probabilistic value.

1.1. *Games*

Given a finite set $N = \{1, 2, \ldots, n\}$ of players, we denote by $\mathcal{N}$ the set of *coalitions*, i.e., of all subsets of N. Then, a n-person cooperative game is a pair (N, v), where v is the *characteristic function* – any real function defined on $\mathcal{N}$ satisfying $v(\emptyset) = 0$. A cooperative game is usually identified with its characteristic function. An important subclass of cooperative games is that of *simple games*, whose characteristic functions take only values 0 and 1, satisfy the condition $v(N) = 1$ and are *monotonic*: if $S \subset T$, then $v(S) \leq v(T)$.

In a simple game v coalition T is *winning* if $v(T) = 1$ and *losing* if $v(T) = 0$. A coalition is *minimal winning* when it is winning but each of its proper subsets is losing. By $\mathcal{W}(v)$ and $\mathcal{M}(v)$ we shall denote, respectively, the set of all winning and of all minimal winning coalitions in v. It is well known that every simple game is uniquely determined by the set of its minimal winning coalitions.

The set of all n -person simple games will be denoted by $\mathcal{P}_n$, and the set of all simple games with finite number of players by $\mathcal{P}^* = \bigcup_{n=1}^{\infty} \mathcal{P}_n$. Similarly, G_n and $\mathcal{G}^* = \bigcup_{n=1}^{\infty} G_n$ will denote corresponding sets of cooperative games. The players will be denoted by small letters $(i, j, k, \ldots)$, and coalitions by capital letters $(S, T, U, \ldots)$. The notation like $\{i,j\}$ will often be abbreviated to ij, and $T \cup \{j\}$ to $T \cup j$.

1.2. *Marginal Contributions and Decisiveness*

The *marginal contribution* of player j to a coalition T (in the game v), denoted by $v_j'(T)$, is the difference $v_j'(T) = v(T) -$

$v(T \setminus j)$ between values of the characteristic function of v on the coalition T with and without player j's participation. In a simple game v, $v'_j(T) = 0$ or $v'_j(T) = 1$ for every j and T.

When $v'_j(T) = 1$, T turns into a losing coalition when j leaves it, and we say that j is *decisive* in T. The set of all coalitions in which player j is decisive (in the game v) will be denoted by $D(j, v)$:

$$D(j, v) = \{S \subset N : v(S) = 1, v(S \setminus j) = 0\} = \{S \subset N : v'_j(S) = 1\}.$$

A player which is not decisive in any coalition (that is, $\forall T \subseteq N$ $v(T) = v(T \setminus j)$) is a *null player*. Players i and j are *interchangeable* in v if for every coalition S including i and j, $v'_i(S) = v'_j(S)$.

1.3. *Precoalitions (A Priori Unions)*

A precoalitions structure P on a player set N is any partition of N into pairwise disjoint subsets:

$$P = \{P_1, P_2, \ldots, P_m\},$$

where $P_1 \cup P_2 \cup \cdots \cup P_m = N$ and for $k \neq l$, $P_k \cap P_l = \emptyset$. The elements of the partition P are called *precoalitions* (or a priori unions). The notion of game with precoalitions originates from Owen (1977); the interpretation is that all players in the same precoalition, say P_k, agree before the play that any coalitions with other players will only be formed collectively – that is, either all players in P_k or none of them can take part. A game (N, v) with imposed precoalitions structure P will be denoted by (N, v, P).

1.4. *Power Indices and Values*

A power index is any function $p : \mathcal{P}^* \to \bigcup_{n=1}^{\infty} \Delta_n$ such that for every n, $p(\mathcal{P}_n) \subset \Delta_n$ (where Δ_n denotes the n-dimensional unit simplex:

$$\Delta_n = \{x = (x_1, x_2, \ldots, x_n) : x_1, \ldots, x_n \geq 0, \sum_{i=1}^{n} x_i = 1\})$$

satisfying

null player property (**NP**): if i is a null player in the game v, then $p_i(v) = 0$, and
equal treatment (**ET**) : if i and j are interchangeable players in v, then $p_i(v) = p_j(v)$.

The "normalization" requirement that power indices of players sum up to unity is sometimes disputed but is natural in our context because individual indices will be explicitly understood as shares when constructing indices with a priori unions. Equal treatment (or even a stronger requirement of symmetry) and the null player property are standard assumptions. For $v \in \mathcal{P}_n$, the coordinates $p_1(v), \ldots, p_n(v)$ of the vector $p(v)$ are interpreted as measures of "power" of individual players; $p_j(v)$ is the *index of player j* in the game v.

Some commonly used indices are obtained by restriction of values to the set of simple games. A *value* for cooperative games is any function $\psi :: \mathcal{G}^* \to \bigcup_{n=1}^{\infty} \mathrm{R}^n$ such that for every n, $\psi(G_n) \subset \mathrm{R}^n$.

1.5. *Owen's Value for Games with Precoalitions*

Owen (1977) defined the Shapley value for games with a priori unions as follows: Given a precoalitions structure $P = \{P_1, P_2, \ldots, P_m\}$ on N, we define for every game $v \in G_n$ and every $k \in \{1, \ldots, m\}$ the "internal game" v_k of the precoalition P_k by

$$v_k(S) = \varphi_k(v_{k|S}) \quad \forall S \subseteq P_k,$$

where φ denotes the Shapley value and $v_{k|S}$ is an m-person game given by

$$v_{k|S}(U) = v\left(\bigcup_{l \in U} P_l\right) \text{ if } k \notin U, \quad v_{k|S}(U) = v\left(\bigcup_{l \in U \setminus k} P_l \cup S\right) \text{ if } k \in U.$$

Then, the values of internal games define values of players in the game with precoalitions structure: $\varphi_j(v, P) = \varphi_j(v_i)\ \forall j \in N$, where i is the number of the precoalition containing player j.

It is worthwhile to note that this construction can in principle be repeated for any value (say, ψ) for cooperative games by simply replacing φ by ψ in all of its stages. This leads to "Owen–ψ" value for games with a priori unions; the Owen–Banzhaf value in Owen (1981) is a prominent example, but also the prenucleolus and its various variants have been generalized to games with precoalitions exactly the same way (see Młodak,

2003). A question remains, of course, how to interpret this construction for values which are not defined as weighted sums of marginal contributions.

2. "COUNTING" POWER INDICES

Most popular power indices (and many values for cooperative games) are defined as weighted sums of marginal contributions of players to coalitions, based on the implicit assumption that the significance of a player in a game depends on the number (and/or set) of coalitions to which the player can effectively contribute. We can say that this amounts to *counting, weighting and normalizing*. Usually, for each player (say, j), either all coalitions in which j is decisive or all minimal winning coalitions to which j belongs, are enumerated, and appropriate positive coefficients (depending only on values of characteristic function on the given coalition and on its subsets) are added over the resulting sets. Then, the numbers obtained for individual players are normalized (= divided by their sum if it differs from unity) to obtain power indices of all players.

Such an algorithm suggests calling power indices computed in this way *counting indices*. Formally, a power index p is counting if it is of the form

$$p_j(v) = \frac{\sum_{S \ni j} \eta_v(S) v'_j(S)}{\sum_{k=1}^{n} \sum_{T \ni k} \eta_v(T) v'_k(T)}, \tag{1}$$

where the coefficients $\eta_v(T)$ are nonnegative and depend only on $v|_T$ – the restriction of v to the coalition T. Equivalently,

$$p_j(v) = \frac{\sum_{S \in D(j,v)} \eta_v(S)}{\sum_{k=1}^{n} \sum_{T \in D(k,v)} \eta_v(T)}.$$

The two best-known indices, *Shapley–Shubik* index φ and *Banzhaf* index b are obviously counting because they are obtained from normalization of symmetric probabilistic values (cf. Section 4). Furthermore, many alternative indices – like the *Deegan–Packel index d* and the *Holler index h*, which take into account only contributions to *minimal* winning coalitions, or the "hybrid" *Johnston* index g – are counting as well.

We do not attempt to provide an exhaustive presentation of power indices here. For the derivation, properties of and intuitions behind the indices named above see Shapley (1953), Banzhaf (1965), Deegan and Packel (1979), Holler (1982), Johnston (1978); a good survey can be found in Freixas and Gambarelli (1997). The table below presents the form of coefficients $\eta_v(\cdot)$ for those indices. Notice that, because of normalization, the index does not change when all coefficients $\eta_v(\cdot)$ are multiplied by a positive constant. This allows us, in particular, to take them equal to unity instead of the traditional $2^{-(n-1)}$ for the Banzhaf index.

Index	Symbol	Form of the coefficients $\eta_v(S)$
Shapley–Shubik	φ	$\frac{(s-1)!(n-s)!}{n!}$
Banzhaf	b	1
Johnston	g	$\frac{1}{\sum_{k\in S} v'_k(S)}$ if $\sum_{k\in S} v'_k(S) > 0$; 0 otherwise
Holler	h	$\min_{k\in S} v'_k(S) (= v(S) - \max_{T\subset S, T\neq S} v(T))$
Deegan–Packel	d	$\frac{\min_{k\in S} v'_k(S)}{s}$

Examples of indices which are not counting include the prenucleolus, obtained from principles clearly different from marginal contributions, and the Napel and Widgren (2001) "strict power" index, whose computation does involve finding the sets $D(\cdot, v)$, but individual indices of players are only computed after comparison of all those sets and elimination of "inferior" players, which is incompatible with the requirement that $\eta_v(S)$ depend only on $v|_S$.

3. EXTENSION TO GAMES WITH PRECOALITIONS STRUCTURE

The method of computing counting power indices suggests a natural way of extending them to games with precoalitions. Take a simple game with a precoalitions structure (N, v, P); according to the interpretation of the structure P, it induces the m-person *external game* $(M, \underline{v})$, where $M = \{1, 2, \ldots, m\}$ and

$$\underline{v}(\underline{U}) = v\left(\bigcup_{l \in \underline{U}} P_l\right) \quad \forall \underline{U} \subseteq \underline{M}.$$

This game, played by precoalitions, is exactly the "quotient game" of Owen (1977). It is a simple game (cf. Proposition 1 below), so any power index – not necessarily counting – can be applied to it to measure "power" of particular precoalitions. However, the question of "power" of individual players from the set N still makes sense in the game with a priori unions, because usually some *proper* subsets of the precoalition P_l would do instead of the whole precoalition in many winning configurations in which "player" l (= precoalition P_l) is decisive in the external game $\underline{v}$. To quote Owen (1977),

> The principal problem, now, lies in determining the division of the total amount among the several members of the *j*th union. It seems reasonable that this division should in some sense reflect the possibilities of the different members of the union – in other words, it should be given by the Shapley value of some game w_j, with the player set S_j, which we must yet define.
> In defining this new game, w_j, it seems natural to take into account, for a given $K \subseteq S_j$, not only the amount $v(K)$ which the members of the coalition K can obtain among themselves, but also the amounts $v\ (K \cup S_p \cup \cdots \cup S_q)$ which they could obtain if they were to defect from S_j and form a coalition with one or more of the remaining unions.

To distinguish between games and players "on different levels", let us use the convention of denoting players in the external game $\underline{v}$ – i.e., the precoalitions – by underlined numbers ($\underline{1}, \underline{2}, \ldots, \underline{m}$) and the coalitions in that game by $\underline{U}, \underline{W}, \ldots$ We have thus $N = \{1, 2, \ldots, n\}$ and $\underline{M} = \{\underline{1}, \underline{2}, \ldots, \underline{m}\}$. Moreover, for any player $j \in N$ denote by $P(j)$ the precoalition to which player j belongs, and by $\underline{i}$ its number: $j \in P(j) = P_{\underline{i}}$.

PROPOSITION 1. *When* (N, v) *is a simple game and* $P = \{P_{\underline{1}}, P_{\underline{2}}, \ldots, P_{\underline{m}}\}$ *is a precoalition structure on* N, *then*

(a) *the external game* $(\underline{M}, \underline{v})$ *is a simple game;*
(b) *for every* $\underline{k} \in \underline{M}$ *and every* $\underline{W} \in D(\underline{k}, \underline{v})$ *the "internal" game* $(P_{\underline{k}}, v_{\underline{k},\underline{W}})$ *defined by*

$$v_{\underline{k},\underline{W}}(S) = v\left(\bigcup_{l \in \underline{W} \setminus \underline{k}} P_{\underline{l}} \cup S\right)$$

is a simple game.

The proof follows directly from the fact that v is a simple game; for part (b) the condition $\underline{W} \in D(\underline{k}, \underline{v})$ assures that $v_{\underline{k},\underline{W}}(\emptyset) = 0$ and $v_{\underline{k},\underline{W}}(P_{\underline{k}}) = 1$. We shall call the game $(P_{\underline{k}}, v_{\underline{k},\underline{W}})$ the *internal game* of precoalition $P_{\underline{k}}$ in the configuration $\underline{W}$.

Proposition 1 allows us to define individual counting indices in the game with a priori unions. The construction resembles that of Owen, but instead of value of one internal game for each $P_{\underline{k}}$ we use power indices of all (simple) games $v_{\underline{k},\underline{W}}$. The algorithm is as follows:

(1) For every $P_{\underline{k}}$ and every $\underline{W} \in D(\underline{k}, \underline{v})$, divide the marginal contribution $v'_{\underline{k}}(\underline{W})$ among all players in $P_{\underline{k}}$ *according to their power indices* in the game $v_{\underline{k},\underline{W}}$, $p.(v_{\underline{k},\underline{W}})$

(2) Weigh each $p_j(v_{\underline{i},\underline{U}})$ with the coefficient $\eta_{\underline{v}}(\underline{U})$ used to compute $p_{\underline{i}}(\underline{v})$. Together with the first step, this gives

$$\eta_v(\underline{W}) = \sum_{j \in P_{\underline{i}}} \eta_{\underline{v}}(\underline{W}) p_j(v_{\underline{i},\underline{W}}) = \sum_{j \in P_{\underline{i}}} \eta_{\underline{v}}(\underline{W}) \frac{\sum_{S \in D(j, v_{\underline{i},\underline{W}})} \eta_{v_{\underline{i},\underline{W}}}(S)}{\sum_{l \in P_{\underline{i}}} \sum_{U \in D(l, v_{\underline{i},\underline{W}})} \eta_{v_{\underline{i},\underline{W}}}(U)}.$$

(3) For every $j \in N$, add all weighted terms to get

$$\tilde{p}_j(v, P) = \sum_{\underline{W} \in D(\underline{i}, \underline{v})} \eta_{\underline{v}}(\underline{W}) p_j(v_{\underline{i},\underline{W}}).$$

(4) Normalize to obtain $p(N, v, P)$:

$$p_j(v, P) = \frac{\tilde{p}_j(v, P)}{\sum_{l=1}^{n} \tilde{p}_l(v, P)} = \frac{\sum_{\underline{W} \in D(\underline{i}, \underline{v})} \eta_{\underline{v}}(\underline{W}) p_j(v_{\underline{i},\underline{W}})}{\sum_{l=1}^{n} \sum_{\underline{U} \in D(\underline{k}, \underline{v})} \eta_{\underline{v}}(\underline{U}) p_l(v_{\underline{k},\underline{U}})}, \quad (2)$$

where $l \in P_{\underline{k}}$.

In other words, the index of every "player" k (=precoalition $P_{\underline{k}}$) in the external game, being a weighted (and normalized) sum of nonzero marginal contributions of $\underline{k}$ to all winning coalitions $\underline{W}$ in $\underline{v}$, is divided among players in $P_{\underline{k}}$ by dividing each of those weighted contributions proportionally to the power indices of individual players in the corresponding internal game ($v_{\underline{k},\underline{W}}$).

This definition implies in particular that indices of games with precoalitions are *consistent* with indices for usual games in the sense proposed by van den Brink and van der Laan (2001) for a related notion of share functions.

PROPOSITION 2. *For each counting power index P and each game (N, v),*

(a) $p(N, v, \{N\}) = p(v)$,
(b) $p(N, v, \{\{1\}, \{2\}, \ldots, \{n\}\}) = p(v)$,
(c) *for every* $\underline{i} \in \underline{M}$, $\sum_{j \in P_{\underline{i}}} p_j(N, v, P) = p_{\underline{i}}(\underline{v})$.

EXAMPLE 1. $N = \{1, 2, 3, 4\}$, $\mathcal{M}(v) = \{12, 13, 14, 234\}$ (as, for instance, in the weighted voting game [40, 20, 20, 20; 55]). For this game,

$$b(v) = \varphi(v) = \frac{1}{6} \cdot (3, 1, 1, 1), \qquad g(v) = \frac{1}{42} \cdot (27, 5, 5, 5),$$
$$h(v) = \frac{1}{9} \cdot (3, 2, 2, 2), \qquad d(v) = \frac{1}{24} \cdot (9, 5, 5, 5).$$

Let $P = \{\{12\}, \{3\}, \{4\}\}$. Then, winning coalitions in $\underline{v}$ are exactly those containing player $\underline{1}$ (i.e., precoalition 12). Therefore players $\underline{2}$ and $\underline{3}$ are null players in the external game, so for every counting index p we have $p_{\underline{2}}(\underline{v}) = p_{\underline{3}}(\underline{v}) = 0$ and, by Proposition 2(c), $p_{\underline{3}}(v, P) = p_{\underline{4}}(v, P) = 0$. By efficiency of the index, $p_{\underline{1}}(\underline{v}) =$ 1 and this is the amount to be shared between players 1 and 2.

The internal games of precoalition 12 are of the form

$$v_{\underline{1},\underline{W}}(\emptyset) = 0, \; v_{\underline{1},\underline{W}}(12) = 1, \quad \text{for every } \underline{W} \in D(\underline{1}, \underline{v}),$$
$$v_{\underline{1},\underline{W}}(1) = v_{\underline{1},\underline{W}}(2) = 0, \quad \text{for } \underline{W} = \{\underline{1}\},$$
$$v_{\underline{1},\underline{W}}(1) = 1, \; v_{\underline{1},\underline{W}}(2) = 0, \quad \text{for } \underline{W} = \{\underline{1}, \underline{2}\} \text{ or } \underline{W} = \{\underline{1}, 3\},$$
$$v_{\underline{1},\underline{W}}(1) = v_{\underline{1},\underline{W}}(2) = 1, \quad \text{for } \underline{W} = \{\underline{M}\},$$

so in each of them either players 1 and 2 are interchangeable, or 2 is a null player. Thus, for every symmetric index p with the null player property,

$$p(v_{\underline{1},\underline{W}}) = \begin{cases} (1,0), & \text{when } \underline{W} = \{\underline{1},\underline{2}\} \text{ or } \underline{W} = \{\underline{1},\underline{3}\}, \\ (\frac{1}{2},\frac{1}{2}), & \text{when } \underline{W} = \{\underline{1}\} \text{ or } \underline{W} = \{\underline{M}\}. \end{cases}$$

These indices are now multiplied by coefficients $\eta_{\underline{v}}(\cdot)$ to obtain

$$\begin{aligned} \tilde{\varphi}_1(v,P) &= \frac{1}{3}\varphi_1(v_{\underline{1},\{\underline{1}\}}) + \frac{1}{6}(\varphi_1(v_{\underline{1},\{\underline{1},\underline{2}\}}) + \varphi_1(v_{\underline{1},\{\underline{1},\underline{3}\}})) + \frac{1}{3}\varphi_1(v_{\underline{1},\underline{M}}) \\ &= \frac{1}{3}\cdot\frac{1}{2} + 2\cdot\frac{1}{6}\cdot 1 + \frac{1}{3}\cdot\frac{1}{2} = \frac{2}{3}, \end{aligned}$$

$$\begin{aligned} \tilde{b}_1(v,P) &= 1\cdot(b_1(v_{\underline{1},\{\underline{1}\}}) + b_1(v_{\underline{1},\{\underline{1},\underline{2}\}}) + b_1(v_{\underline{1},\{\underline{1},\underline{3}\}}) + b_1(v_{\underline{1},\underline{M}})) \\ &= \frac{1}{2} + 2\cdot 1 + \frac{1}{2} = 3, \end{aligned}$$

and

$$\tilde{\varphi}_2(v,P) = \frac{1}{3}\cdot\frac{1}{2} + 2\cdot\frac{1}{6}\cdot 0 + \frac{1}{3}\cdot\frac{1}{2} = \frac{1}{3}, \quad \tilde{b}_2(v,P) = \frac{1}{2} + 2\cdot 0 + \frac{1}{2} = 1,$$

so finally

$$\varphi(v,P) = \frac{1}{3}\cdot(2,1,0,0), \quad b(v,P) = \frac{1}{4}\cdot(3,1,0,0).$$

Similar computations give $g(v,P) = b(v,P) = \frac{1}{4}\cdot(3,1,0,0)$.

Since $\{\underline{1}\}$ is the only minimal winning coalition in $\underline{v}$, for every other coalition $\underline{W}$ it must be that $\min_{\underline{k}\in\underline{W}} v'_{\underline{k}}(\underline{W}) = 0$. Therefore the only nonzero term in Holler and Deegan–Packel indices of players 1 and 2 is $p_j(v_{\underline{1},\{\underline{1}\}})$, $j = 1,2$, which is equal for both players in this precoalition, and so

$$h(v,P) = d(v,P) = \frac{1}{2}\cdot(1,1,0,0).$$

EXAMPLE 2. Let $N = \{1,2,3,4\}$ and $\mathcal{M}(v) = \{12,13,14\}$ (as in the weighted voting game [46, 18, 18, 18; 55]. This example is borrowed from van den Brink and van der Laan (2001)). In the game (N, v) we have

$$b(v) = \frac{1}{10}\cdot(7,1,1,1), \qquad \varphi(v) = \frac{1}{12}\cdot(9,1,1,1),$$

$$h(v) = d(v) = \frac{1}{6}\cdot(3,1,1,1), \qquad g(v) = \frac{1}{14}\cdot(11,1,1,1).$$

(2a). For precoalitions structure $P = \{\{12\},\{3\},\{4\}\}$, as in Example 1, player $\underline{1}$ in the external game is decisive in every coalition to which he belongs, and players $\underline{2}$ and $\underline{3}$ are null players, so $p_{\underline{3}}(v,P) = p_{\underline{4}}(v,P) = 0$ for every counting index p. Internal games of precoalition 12 are given by

$$v_{\underline{1},\underline{W}}(1) = v_{\underline{1},\underline{W}}(2) = 0, \quad \text{for } \underline{W} = \{\underline{1}\},$$

$$v_{\underline{1},\underline{W}}(1) = 1, \quad v_{\underline{1},\underline{W}}(2) = 0, \quad \text{for every } \underline{W} \in D(\underline{1},\underline{v}),\ \underline{W} \neq \{\underline{1}\}$$

(and $v_{\underline{1},\underline{W}}(\varnothing) = 0$, $v_{\underline{1},\underline{W}}(12) = 1\ \forall \underline{W} \in D(\underline{1},\underline{v})$), so for every index p with the properties **ET** and **NP**

$$p(v_{\underline{1},\underline{W}}) = \begin{cases} \left(\frac{1}{2},\frac{1}{2}\right) & \text{when } \underline{W} = \{\underline{1}\}, \\ (1,0) & \text{when } \underline{W} \neq \{\underline{1}\}. \end{cases}$$

Therefore

$$\varphi_1(v,P) = \frac{1}{3}\cdot\frac{1}{2} + 2\cdot\frac{1}{6}\cdot 1 + \frac{1}{3}\cdot 1 = \frac{5}{6}, \quad \varphi_2(v,P) = \frac{1}{3}\cdot\frac{1}{2} = \frac{1}{6},$$

$$\tilde{b}_1(v,P) = \tilde{g}_1(v,P) = 1\cdot\left(\frac{1}{2} + 2\cdot 1 + 1\right) = \frac{7}{2}, \quad \tilde{b}_2(v,P) = \tilde{g}_2(v,P) = \frac{1}{2},$$

and so

$$\varphi(v,P) = \frac{1}{6}\cdot(5,1,0,0), \quad b(v,P) = g(v,P) = \frac{1}{8}\cdot(7,1,0,0),$$

while for Holler and Deegan–Packel indices, as in Example 1,

$$h(v,P) = d(v,P) = \frac{1}{2}\cdot(1,1,0,0).$$

(2b). With precoalitions structure $P = \{\{1\},\{234\}\}$, both the external game $\underline{v}$ and the unique internal game $u(= v_{\underline{2},\underline{M}})$ are symmetric, so for every index p

$$p_{\underline{1}}(\underline{v}) = p_{\underline{2}}(\underline{v}) = \frac{1}{2} \quad \text{and} \quad p_2(u) = p_3(u) = p_4(u) = \frac{1}{3}.$$

Thus $p(N,v,P) = \frac{1}{6}\cdot(3,1,1,1)$ for any counting index p.

Examples 1 and 2a show that Holler and Deegan–Packel indices can behave in a quite strange way in games endowed with precoalitions structure. However, this observed equating of players within precoalitions results rather from the definition of these indices themselves than from our method of extending indices to games with a priori unions. Consider a precoalition $P_{\underline{k}}$

of a "strong" player and some "weaker" players, being a minimal winning coalition in v (and thus also in $\underline{v}$). By entering it, the stronger player loses all "advantage" resulting from ampler possibilities of forming alternative coalitions. He can be decisive in many coalitions which do not split precaolitions different from $\underline{k}$ and may even be minimal winning in v, but all those contributions will be counted with zero weights by Holler and Deegan–Packel indices, because the resulting winning coalitions in the external game in which $\underline{k}$ is decisive are not minimal winning.

4. INDICES DERIVED FROM SYMMETRIC PROBABILISTIC VALUES

A *symmetric probabilistic value* (SPV) ψ assigns to any n-person cooperative game $v \in G_n$ a vector $(\psi_1(v), \ldots, \psi_n(v))$ with components given by

$$\psi_j(v) = \sum_{T \subset N} c_n(T) v_j'(T), \tag{3}$$

where $c_n(T)$ are some nonnegative coefficients fulfilling

$$\text{(a) } \#S = \#T \Rightarrow c_n(S) = c_n(T),$$

$$\text{(b) } \forall i \in N \quad \sum_{T \ni i} c_n(T) = 1$$

(hence the terms "symmetric" and "probabilistic"), and thus depending only on the number of players, n, and on $\#T$, but not on the game v.

It is clear that every SPV ψ of the form (3) leads to a counting power index p^{ψ} on $\mathcal{P}^*$, given for $v \in \mathcal{P}_n$ by the formulae

$$\tilde{p}_j^{\psi}(v) = \sum_{T \subset N} c_n(T) v_j'(T) = \sum_{U \in D(j,v)} c_n(U) = \psi_j(v),$$

$$p_j^{\psi}(v) = \frac{\tilde{p}_j^{\psi}(v)}{\sum_{k=1}^{n} \tilde{p}_k^{\psi}(v)} = \frac{\psi_j(v)}{\sum_{k=1}^{n} \psi_k(v)}$$

(which amounts to restriction of ψ to $\mathcal{P}^*$ and subsequent normalization). This index, of course, can be extended to games with precoalitions using the method described in Section 3.

On the other hand, we know from Section 1 that every value on $\mathcal{G}^*$ (not necessarily symmetric or probabilistic) can itself be extended to games with precoalitions structure using Owen (1977) method. The obtained "Owen-ψ" value $\mathrm{O}\psi$ can then be restricted to simple games with a priori unions and normalized to give a new power index $p^{\mathrm{O}\psi}$:

$$p_j^{\mathrm{O}\psi}(v, P) = \frac{\mathrm{O}\psi_j(v, P)}{\mathrm{O}\psi_1(v, P) + \cdots + \mathrm{O}\psi_n(v, P)}.$$

A comparison of the two indices may be of interest; it turns out that they usually differ.

THEOREM 1. *Let ψ be a probabilistic symmetric value, and let p^{ψ} and $p^{\mathrm{O}\psi}$ be power indices for games with precoalitions, derived from ψ as described above. Then $p^{\psi} = p^{\mathrm{O}\psi}$ if and only if ψ is the Shapley value.*

Proof. Let (N, v, P) be any n-person game with precoalitions structure. For every $j \in N$, denote by $a(j)$ the cardinality of the precoalition $P_{\underline{i}}$ containing player j. Given a SPV ψ, denote by $c_{t,l}$ the common value of coefficients $c_l(T)$ in the value ψ for all t-person coalitions T in a l-person game. Then

$$\mathrm{O}\psi_j(N, v, P) = \psi_j(v_{\underline{i}}) = \sum_{S \ni j, S \subseteq P_{\underline{i}}} c_{s,a(j)}(v_{\underline{i}}(S) - v_{\underline{i}}(S \setminus j)).$$

After appropriate substitutions and changing the order of summation, the right-hand side simplifies to the form

$$\sum_{\underline{W} \subseteq \underline{M} \setminus \underline{i}} c_{w+1,m} \sum_{S \subseteq P_{\underline{i}}} c_{s,a(j)} \left[v\left(\bigcup \underline{W} \cup S\right) - v\left(\bigcup \underline{W} \cup S \setminus j\right)\right],$$

where $w = \#\underline{W}$. When v is a simple game, the expression in square brackets is always equal to zero or unity, so

$$\mathrm{O}\psi_j(N, v, P) = \sum_{S, \underline{W}: v(\bigcup \underline{W} \cup S) > v(\bigcup \underline{W} \cup S \setminus j)} c_{s,a(j)} c_{w+1.m} \tag{4}$$

On the other hand, for the index p^{ψ} we have

$$\tilde{p}_j^{\psi}(N,v,P) = \sum_{\underline{W}\in D(\underline{i},\underline{v})} c_{w,m}\, p_j^{\psi}(v_{\underline{i},\underline{W}}) = \sum_{\underline{W}\in D(\underline{i},\underline{v})} c_{w,m} \frac{\psi_j(v_{\underline{i},\underline{W}})}{\sum_{k\in P_{\underline{i}}} \psi_k(v_{\underline{i},\underline{W}})}$$

$$= \sum_{\underline{W}\in D(\underline{i},\underline{v})} c_{w,m} \frac{\sum_{S\in D(j,v_{\underline{i},\underline{W}})} c_{s,a(j)}}{\sum_{k\in P_{\underline{i}}} \sum_{T\in D(k,v_{\underline{k},\underline{W}})} c_{t,a(j)}}. \tag{5}$$

When ψ is the Shapley value, the denominator is always equal to $v_{\underline{i},\underline{W}}(P_{\underline{i}})$. Therefore, since for $\underline{W} \in D(\underline{i}, \underline{v})$ the game $v_{\underline{i},\underline{W}}$ is simple, in the case $\psi = \phi$ all denominators are equal to 1 and thus

$$O\psi_j(N,v,P) = \sum_{\underline{W}\in D(\underline{i},\underline{v})} c_{w,m} \sum_{S\in D(j,v_{\underline{i},\underline{W}})} c_{s,a(j)} = \tilde{p}_j^{\psi}(N,v,P) \quad \forall j \in N.$$

Moreover, successive application of efficiency of the Shapley value to all games $v_{\underline{k},\underline{W}}$ and to the external game gives $\sum_{j\in N} \tilde{p}_j^{\psi}(N, v, P) = 1$. Therefore

$$p_j^{O\varphi}(N, v, P) = p_j^{\varphi}(N, v, P) \quad \forall j \in N,$$

so the indices φ and $O\varphi$ are equal on all games with precoalition structures.

To show that for all other (inefficient) SPVs the equalities following (5) do not hold, consider the game (N, v, P) with $N = \{1, 2, \ldots, n\}$, $\mathcal{M}(v) =$ all coalitions containing player 1 and exactly t other players, and $P = \{1, N\backslash 1\}$, which is a generalized version of the game in Example 2b. Again, for each index p^{ψ} derived from some SPV (in fact, for each counting index) equal treatment and consistency enforce

$$p^{\psi}(v, P) = \frac{1}{2(n-1)} \cdot (n-1, 1, \ldots, 1).$$

But by (4)

$$\tilde{p}_1^{O\psi}(v, P) = O\psi_1(v, P) = c_{2,2} \cdot c_{1,1},$$

$$\tilde{p}_j^{O\psi}(v, P) = O\psi_j(v, P) = c_{2,2} \cdot \binom{n-2}{t-1} \cdot c_{t,n-1}, \text{ for each } j > 1,$$

because there are $\binom{n-2}{t-1}$ coalitions in which player j is decisive in the internal game $v_{\underline{2},\underline{M}}$. Together with the preceding equation, this implies

$$c_{2,2} \cdot c_{1,1} = (n-1) \cdot c_{2,2} \cdot \binom{n-2}{t-1} \cdot c_{t,n-1}.$$

Since $c_{1,1} = 1$ (as the index of the unique player in a one-person simple game) and $c_{2,2} \neq 0$, this gives

$$c_{t,n-1} = \frac{(t-1)!(n-t-1)!}{(n-1)!},$$

which is exactly the coefficient in the Shapley value. □

Notice that the assumption of the value ψ being probabilistic is not essential for constructing the index p^{ψ}: because of normalization, multiplying all coefficients by a positive constant would lead to the same index. We use it rather for terminological reasons since there is no established name for values being weighted sums (with nonnegative and symmetric weights) of marginal contributions. Without the "probability" assumption, any values proportional to the Shapley value can substitute φ in Theorem 1.

ACKNOWLEDGEMENT

This paper has been supported by a KBN (Polish State Committee for Scientific Research) project 5 H02B 001 21 "Group decisions – structure, power indices and individual rights".

REFERENCES

Banzhaf, J.F. (1965), Weighted voting does not work: A mathematical analysis, *Rutgers Law Review* 19, 317–343.

Deegan, J. and Packel, E.W. (1979), A new index of power for simple N-person games, *International Journal of Game Theory* 7, pp. 113–123.

Freixas, J. and Gambarelli, G. (1997), Common internal properties among power indices, *Control and Cybernetics* 26, 591–603.

Holler, M.J. (1982), Forming coalitions and measuring voting power, *Political Studies* 30(2), 266–271.

Johnston, R.J. (1978), On the measurement of power: some reactions to Laver, *Environment and Planning* 10A, 907–914.

Młodak, A. (2003), Three additive solutions of cooperative games with a priori unions, *Applicationes Mathematicae* 30, 69–87.

Napel, S. and Widgren, M. (2001), Inferior players in simple games, *International Journal of Game Theory* 30, 209–220.

Owen, G. (1977), Values of games with a priori unions, in R. Henn and O. Moeschlin (eds), *Mathematical Economics and Game Theory: Essays in Honor of Oskar Morgenstern* (Springer-Verlag).

Owen, G. (1981), Modification of the Banzhaf–Coleman index for games with a priori unions, in M.J. Holler (ed), *Power, Voting and Voting Power* (Physica-Verlag).

Shapley, L.S. (1953), A value for *N*-person games, in H. Kuhn and A.W. Tucker (eds), *Contributions to the Theory of Games*, Vol. 2. (Princeton: University Press.

Van den Brink, R. and van der Laan, G. (2001), A class of consistent share functions for games in coalition structure. Discussion Paper 33, CentER, Tilburg (to appear in *Games and Economic Behavior*).

Address for correspondence: Marcin Malawski, Instytut Podstaw Informatyki PAN, Ordona 21, 01-237 Warszawa, Poland. Tel.: +48-22-836-3709; Fax: +48-22-837-6564; E-mail: malawski@ipipan.waw.pl

ROSS CRESSMAN, JÓZSEF GARAY, ANTONINO SCARELLI and ZOLTÁN VARGA

THE DYNAMIC STABILITY OF COALITIONIST BEHAVIOUR FOR TWO-STRATEGY BIMATRIX GAMES*

ABSTRACT. In this paper replicator dynamics are introduced to describe the propagation of coalitionist behaviour in conflicts given by a two-strategy bimatrix games. In the proposed approach non-coalitionists play either Nash strategies or choose one of two pure strategies. In the first case it is proved that non-coalitionists are asymptotically eliminated. In the second case coalitionists can propagate without eliminating all non-coalitionists.

KEY WORDS: bimatrix game, replicator dynamics, imitation, coalition

1. INTRODUCTION

Evolutionary game theory applied to bimatrix games has traditionally been restricted to predicting the players' behaviour in the context of a non-cooperative game. The purpose of this paper is to investigate continuous dynamic models that describe the evolution of the frequency for cooperative coalitionist behaviour relative to that of non-cooperative behaviour in pair-wise conflicts between agents of two populations that are given as a bimatrix game.

The basic idea is to adapt the so-called replicator dynamics describing the time variation of certain behaviour types in a population (Hofbauer and Sigmund, 1998; Cressman, 1992). In the biological evolutionary game context, where coalition formation has received little attention, replication means that "like begets like": a behaviour type is passed down from parent to offspring. Adapting this idea sketched in Weibull (1995), see also Hofbauer and Sigmund (1998), our model can also be applied to

* The research was supported by the Hungarian National Research Fund (OTKA) No. T037271. The final version was completed while one of the authors (R.C.) was a Fellow at the Collegium Budapest.

Theory and Decision **56:** 141–152, 2004.

the social evolution of behaviour types (strategies) in a population of pair-wise interacting agents, such as when individuals represent firms or some other social or economic units. The replication of carriers of a behaviour type can then be thought of as resulting from imitation. An agent is ready to review his strategy at a certain rate depending on the success of another strategy. In the corresponding replicator dynamics, the time rate of a strategy's relative frequency change is positive if this strategy performs better than the average, and negative in the opposite case.

Originally the replicator dynamics was introduced for the study of game-theoretical conflicts in biological evolutionary processes. This is based on the Darwinian theory where the competitiveness of behaviour types is the key issue since these types will increase in relative frequency if their gain (fitness) is greater than the mean gain (fitness) of the whole population. The aim of the present paper is then to study whether replicator dynamics corresponding to the theory of cooperative games can describe the propagation of coalitionist behaviour.

We propose dynamic models where in each population of agents the basic behaviour types are: coalitionist and non-coalitionist. In the *monomorphic dynamic model*, based on a two-strategy two-person (bimatrix) game, in the pair-wise conflict of coalitionists a two-person coalition is formed, and always the Nash equilibrium strategies are played if at least one of the agents is non-coalitionist (Owen, 1968, 1995). In the *polymorphic dynamic model*, the pair-wise conflict of two coalitionists is again modelled by a two-person coalition game, but if at least one of the agents is non-coalitionist, both use one of the two fixed pure strategies. In each model, the imputation is a fixed parameter. The asymptotic behaviour of these models is studied and a numerical illustration is also given.

2. BIMATRIX GAMES AND TWO-PERSON COALITIONS

Let us consider a bimatrix game with pay-off matrices

$$\mathbf{A} := \begin{pmatrix} a_{11} & a_{12} \\ a_{21} & a_{22} \end{pmatrix} \quad \text{and} \quad \mathbf{B} := \begin{pmatrix} b_{11} & b_{12} \\ b_{21} & b_{22} \end{pmatrix}.$$

The state space, i.e. the Cartesian product of mixed strategy sets, is then $S_2 \times S_2$ where $S_2 = \{p = (p_1, p_2) : p_1 + p_2 = 1, p_1 \geq 0, p_2 \geq 0\}$. First we look for an interior (totally mixed) Nash equilibrium, i.e., a strategy pair $(\mathbf{p}^*, \mathbf{q}^*) \in \text{int } S_2 \times S_2$ with $\mathbf{p}^*\mathbf{A}\mathbf{q}^* \geq \mathbf{p}\mathbf{A}\mathbf{q}^*$ and $\mathbf{q}^*\mathbf{B}\mathbf{p}^* \geq \mathbf{q}\mathbf{B}\mathbf{p}^*$ for all $(\mathbf{p}, \mathbf{q}) \in S_2 \times S_2$. Throughout the paper, we will assume that $sign(a_{22} - a_{12}) = sign(a_{11} - a_{21}) \neq 0$ and $sign(b_{22} - b_{12}) = sign(b_{11} - b_{21}) \neq 0$. Then the unique interior Nash equilibrium is $(\mathbf{p}^*, \mathbf{q}^*)$ with

$$\mathbf{p}^* := \left(\frac{b_{22} - b_{12}}{b_{11} - b_{12} - b_{21} + b_{22}}, \frac{b_{11} - b_{21}}{b_{11} - b_{12} - b_{21} + b_{22}}\right), \quad (2.1)$$

and

$$\mathbf{q}^* := \left(\frac{a_{22} - a_{12}}{a_{11} - a_{12} - a_{21} + a_{22}}, \frac{a_{11} - a_{21}}{a_{11} - a_{12} - a_{21} + a_{22}}\right). \quad (2.2)$$

Let us consider now the two-person co-operative (coalition) game corresponding to the above bimatrix game. In addition we shall assume that no entries of the matrix $\mathbf{A} + \mathbf{B}^{\mathrm{T}}$ are equal. Then it is easy to see that the coalition game is *NE-essential*[1] in the sense that

$$\max_{\mathbf{p},\mathbf{q} \in S_2} \mathbf{p}(\mathbf{A} + \mathbf{B}^{\mathrm{T}})\mathbf{q} = \max_{i,j \in \{1,2\}} (a_{ij} + b_{ji}) > \mathbf{p}^*(\mathbf{A} + \mathbf{B}^{\mathrm{T}})\mathbf{q}^*.$$

If the strictly maximal entry of matrix $\mathbf{A} + \mathbf{B}^{\mathrm{T}}$ is $a_{kl} + b_{lk}$, then the strategy pair providing the maximum pay-off to the two-person coalition is the corresponding pure strategy pair denoted by $(\mathbf{e}_k, \mathbf{e}_l)$. For a given NE-essential game a pair $(u, v) \in \mathbf{R}^2$ will be called an *NE-imputation*, if

$$u + v = \max_{\mathbf{p},\mathbf{q} \in S_2} \mathbf{p}(\mathbf{A} + \mathbf{B}^{\mathrm{T}})\mathbf{q} \quad \text{and} \quad \mathbf{p}^*\mathbf{A}\mathbf{q}^* \leq \mathrm{u}, \mathbf{q}^*\mathbf{B}\mathbf{p}^* \leq \mathrm{v},$$

where at least one of the inequalities is strict, and (u, v) is said to be a *strict NE-imputation* if both inequalities are strict.

3. MONOMORPHIC REPLICATOR DYNAMICS

Let us consider populations I and II consisting only of non-cooperative individuals playing their corresponding interior Nash equilibrium strategies. In evolutionary game theory the populations in such a situation are called monomorphic, since all players in the same population have the same behavioural phenotype. In this section we shall consider whether the coalitionist behaviour would propagate in such populations.

Let us assume that there are two possible behaviour types for the agents: coalitionist and non-coalitionist. In a pair-wise conflict of two coalitionists a two-person coalition is formed with the fixed NE-imputation (u, v). If at least one of the agents is non-coalitionist, then both agents play their respective NE strategy. It is assumed that the relative rate of change in frequency of a behaviour in each population is equal to the difference between the expected payoff to the given behaviour type and the average payoff over the corresponding population (social or economic unit). Denote by x_1 and x_2 the frequencies of coalitionist and non-coalitionist behaviours in population I, respectively, and put $\mathbf{x} = (x_1, x_2)$. The meaning of $\mathbf{y} = (y_1, y_2)$ is similar for population II. Then the corresponding replicator dynamics is

$$\begin{aligned} \dot{x}_1 &= x_1[uy_1 + \mathbf{p}^*\mathbf{A}\mathbf{q}^* y_2 - V(\mathbf{x}, \mathbf{y})], \\ \dot{x}_2 &= x_2[\mathbf{p}^*\mathbf{A}\mathbf{q}^* - V(\mathbf{x}, \mathbf{y})], \\ \dot{y}_1 &= y_1[vx_1 + \mathbf{q}^*\mathbf{B}\mathbf{p}^* x_2 - W(\mathbf{x}, \mathbf{y})], \\ \dot{y}_2 &= y_2[\mathbf{q}^*\mathbf{B}\mathbf{p}^* - W(\mathbf{x}, \mathbf{y})] \end{aligned} \tag{3.1}$$

where $V(\mathbf{x}, \mathbf{y}) = ux_1y_1 + \mathbf{p}^*\mathbf{A}\mathbf{q}^*(x_1y_2 + x_2)$ and $W(\mathbf{x}, \mathbf{y}) = vx_1y_1 + \mathbf{q}^*\mathbf{B}\mathbf{p}^*(y_1x_2 + y_2)$ are the average payoffs of the players in population I and II, respectively. The set $S_2 \times S_2$ and its "faces" are positively invariant. A rest point must satisfy the following equalities

$$\begin{aligned} (\mathbf{p}^*\mathbf{A}\mathbf{q}^* - u)y_1x_1x_2 &= 0, \\ (\mathbf{q}^*\mathbf{B}\mathbf{p}^* - v)x_1y_1y_2 &= 0. \end{aligned}$$

Thus, since at least one of the inequalities $\mathbf{p}^*\mathbf{A}\mathbf{q}^* \leq u$, $\mathbf{q}^*\mathbf{B}\mathbf{p}^* \leq v$ is strict, there is no interior rest point.

THEOREM 1. *For any strict NE-imputation* (u, v) , *the coalitionist state* ((1,0), (1,0)) *is globally asymptotically stable for dynamics* (3.1).

Proof. Observe that

$$\begin{aligned}\mathbf{p}^*\mathbf{A}\mathbf{q}^* - V(\mathbf{x},\mathbf{y}) &= \mathbf{p}^*\mathbf{A}\mathbf{q}^* - ux_1y_1 - \mathbf{p}^*\mathbf{A}\mathbf{q}^*(x_1y_2 + x_2)\\ &= \mathbf{p}^*\mathbf{A}\mathbf{q}^*(1 - (x_1y_2 + x_2)) - ux_1y_1\\ &= (\mathbf{p}^*\mathbf{A}\mathbf{q}^* - u)x_1y_1,\end{aligned}$$

and similarly

$$\mathbf{q}^*\mathbf{B}\mathbf{p}^* - W(\mathbf{x},\mathbf{y}) = (\mathbf{q}^*\mathbf{B}\mathbf{p}^* - v)x_1y_1.$$

Thus $\mathbf{p}^*\mathbf{A}\mathbf{q}^* < u$ and $\mathbf{q}^*\mathbf{B}\mathbf{p}^* < v$ imply $\dot{x}_2 < 0$ and $\dot{y}_2 < 0$, respectively. Therefore, the coalitionist agents with imputation greater than the corresponding NE payoff will win in the sense that, in the long term, non-coalitionists are eliminated from the corresponding population. □

4. POLYMORPHIC REPLICATOR DYNAMICS

Let us consider populations I and II where pairs of individuals who are not involved in a cooperative game must play one of their pure strategies. In evolutionary game theory, the populations in such a situation are called polymorphic, since within each population several behavioural phenotypes may be present at a given time. In this section we shall study whether the coalitionist behaviour would propagate in such populations.

Let us start out from an NE-essential bimatrix game as given above, and fix a NE-imputation (u, v) .

We again assume that, in a pair-wise conflict of two coalitionists, a two-person coalition is formed with imputation (u, v). However, unlike the monomorphic model, if at least one of the agents is a non-coalitionist, then fixed pure strategies ($\mathbf{e}_1$ or $\mathbf{e}_2$) are played. Denote by x_1 and x_2 the relative frequencies of coalitionists of population I playing $\mathbf{e}_1$ and $\mathbf{e}_2$, respectively against non-coalitionists of population II. Furthermore, let x_3 and x_4 be the relative frequencies of non-coalitionists of population I playing $\mathbf{e}_1$ and $\mathbf{e}_2$, respectively. We use similar notation y_1, y_2, y_3, y_4 for population II. This situation can be described by a 4-player bimatrix game with the following partitioned payoff matrices

$$\bar{\mathbf{A}} := \begin{pmatrix} \mathbf{U} & \mathbf{A} \\ \mathbf{A} & \mathbf{A} \end{pmatrix} \in \mathbf{R}^{4\times 4}, \qquad \overline{\mathbf{B}} := \begin{pmatrix} \mathbf{V} & \mathbf{B} \\ \mathbf{B} & \mathbf{B} \end{pmatrix} \in \mathbf{R}^{4\times 4},$$

where

$$\mathbf{U} := \begin{pmatrix} u & u \\ u & u \end{pmatrix} \quad \text{and} \quad \mathbf{V} := \begin{pmatrix} v & v \\ v & v \end{pmatrix}.$$

For this 4-person game the corresponding polymorphic replicator dynamics is

$$\begin{aligned}
\dot{x}_1 &= x_1[u(y_1 + y_2) + e_1\mathbf{A}(y_3, y_4) - V(\mathbf{x}, \mathbf{y})], \\
\dot{x}_2 &= x_2[u(y_1 + y_2) + e_2\mathbf{A}(y_3, y_4) - V(\mathbf{x}, \mathbf{y})], \\
\dot{x}_3 &= x_3[e_1\mathbf{A}(y_1 + y_3, y_2 + y_4) - V(\mathbf{x}, \mathbf{y})], \\
\dot{x}_4 &= x_4[e_2\mathbf{A}(y_1 + y_3, y_2 + y_4) - V(\mathbf{x}, \mathbf{y})], \\
\dot{y}_1 &= y_1[v(x_1 + x_2) + e_1\mathbf{B}(x_3, x_4) - W(\mathbf{x}, \mathbf{y})], \\
\dot{y}_2 &= y_2[v(x_1 + x_2) + e_2\mathbf{B}(x_3, x_4) - W(\mathbf{x}, \mathbf{y})], \\
\dot{y}_3 &= y_3[e_1\mathbf{B}(x_1 + x_3, x_2 + x_4) - W(\mathbf{x}, \mathbf{y})], \\
\dot{y}_4 &= y_4[e_2\mathbf{B}(x_1 + x_3, x_2 + x_4) - W(\mathbf{x}, \mathbf{y})],
\end{aligned} \tag{4.1}$$

where

$$\begin{aligned}
V(\mathbf{x}, \mathbf{z}) = u(x_1 + x_2)(y_1 + y_2) + (x_1, x_2)\mathbf{A}(y_3, y_4) + \\
+ (x_3, x_4)\mathbf{A}(y_1 + y_3, y_2 + y_4)
\end{aligned}$$

and

$$\begin{aligned}
W(\mathbf{x}, \mathbf{z}) = v(x_1 + x_2)(y_1 + y_2) + (y_1, y_2)\mathbf{B}(x_3, x_4) + \\
+ (y_3, y_4)\mathbf{B}(x_1 + x_3, x_2 + x_4)
\end{aligned}$$

denote the average payoffs in populations I and II, respectively.

The set $S_4 \times S_4$ and its "faces" are positively invariant under the above dynamics. It is easy to see that the "coalitionist face", $C := \{(\mathbf{x}, \mathbf{y}) \in (S_4 \times S_4) | x_3 = x_4 = 0 \text{ and } y_3 = y_4 = 0\}$ consists only of rest points of dynamics (4.1). Since under our conditions the initial bimatrix game has a unique interior NE, it is easily seen that the unique non-trivial rest point of dynamics (4.1) in the "non-coalitionist face" $D := \{(\mathbf{x}, \mathbf{y}) \in (S_4 \times S_4) | x_1 = x_2 = 0 \text{ and } y_1 = y_2 = 0\}$ is $(\mathbf{x}^*, \mathbf{y}^*)$ with

$$\mathbf{x}^* = \left(0, 0, \frac{b_{22} - b_{12}}{b_{11} - b_{12} - b_{21} + b_{22}}, \frac{b_{11} - b_{21}}{b_{11} - b_{12} - b_{21} + b_{22}}\right),$$

$$\mathbf{y}^* = \left(0, 0, \frac{a_{22} - a_{12}}{a_{11} - a_{12} - a_{21} + a_{22}}, \frac{a_{11} - a_{21}}{a_{11} - a_{12} - a_{21} + a_{22}}\right).$$

Observe that the non-zero coordinates provide the Nash equilibrium of the original bimatrix game of the non-coalitionist agents of the different populations. We notice that, within the non-coalitionist face, either the interior Nash equilibrium is neutrally stable with periodic orbits or two of the vertices are locally asymptotically stable (Hofbauer and Sigmund, 1998).

Now we shall find conditions implying that the interior of the "non-coalitionist face" is a repellor.

THEOREM 2. *Suppose that for each payoff matrix the arithmetic mean of each column is not greater than the corresponding NE-imputation component:*

$$\begin{aligned} u &\geq \frac{1}{2}(a_{11} + a_{21}), \quad u \geq \frac{1}{2}(a_{12} + a_{22}), \\ v &\geq \frac{1}{2}(b_{11} + b_{21}), \quad v \geq \frac{1}{2}(b_{12} + b_{22}) \end{aligned} \tag{4.2}$$

with at least one strict inequality.

Then the interior of D is a repellor (i.e. no trajectory in the interior of $S_4 \times S_4$ has an ω-limit point in the interior of D).

Proof. Define a function

$$L : \operatorname{int}(S_4 \times S_4) \to \mathbf{R},$$

$$L(\mathbf{x}, \mathbf{y}) := \ln x_1 + \ln x_2 - \ln x_3 - \ln x_4 + \ln y_1 + \ln y_2 - \ln y_3 - \ln y_4.$$

Then along any trajectory of dynamics (4.1) starting from the interior of the strategy set, the function L is strictly increasing. Indeed, the derivative of L with respect to the considered dynamics is

$$DL(\mathbf{x},\mathbf{y}) = (1,1)[\mathbf{U}-\mathbf{A}](y_1,y_2) + (1,1)[\mathbf{V}-\mathbf{B}](x_1,x_2)$$
$$(\mathbf{x},\mathbf{y}) \in \text{int}(S_4 \times S_4).$$

Condition (4.2) on the imputation implies that for all $y_1 > 0$, $y_2 > 0$, $x_1 > 0$ and $x_2 > 0$ we have

$$(2u - a_{11} - a_{21})y_1 + (2u - a_{12} - a_{22})y_2 + (2v - b_{11} - b_{21}) x_1 + (2v - b_{12} - b_{22})x_2 > 0$$

Hence $DL(\mathbf{x},\mathbf{y}) > 0$, implying that function L is strictly increasing along each trajectory starting from the interior. Since the function L approaches $-\infty$ near any point of the interior of D, the latter is a repellor. □

Remark 1. The condition (4.2) of Theorem 2 can be related to the concept of dominance used by Akin (1980) for symmetric two-person matrix games.

Following Akin (1980) and Akin and Hofbauer (1982) we can prove

THEOREM 3. *Under the conditions of Theorem 2 we have* $x_3(t)x_4(t)y_3(t)y_4(t) \to 0$ as $t \to +\infty$ *for any trajectory in the interior of* $S_4 \times S_4$.

Proof. Let (x,y) be a solution of dynamics (4.1) with $(x(0),y(0)) \in \text{int}(S_4 \times S_4)$. Since the "faces" of $S_4 \times S_4$ are invariant, so is $\text{int}(S_4 \times S_4)$. By the proof of Theorem 2, $L(x(t),y(t))$ is strictly increasing. First we show that

$$\lim_{t\to+\infty} L(x(t),y(t)) = +\infty. \tag{4.4}$$

Suppose the contrary: for some $c \in \mathbf{R}$ we have $\lim_{t\to+\infty} L(x(t), y(t)) = c$. This implies $\lim_{t\to+\infty} d/dt\ L(x(t),y(t)) = 0$. By the compactness of $S_4 \times S_4$ the solution (x,y) has an ω-limit point $(x^0,y^0) \in S_4 \times S_4$. Then, for an appropriate $t_n \to +\infty$, we have $\lim_{n\to+\infty} d/dt\ L(x(t_n),y(t_n)) = 0$. By inequality (4.2), the latter implies $x_1^0 = x_2^0 = y_1^0 = y_2^0 = 0$. Hence $\lim_{n\to+\infty} L(x(t_n),y(t_n)) = +\infty$ which is a contradiction to the supposed finite limit.

Now from (4.4) we conclude

$$\lim_{t\to+\infty} \frac{x_1(t)x_2(t)}{x_3(t)x_4(t)} \frac{y_1(t)y_2(t)}{y_3(t)y_4(t)} = +\infty.$$

Since the numerator in the quotient is bounded, we have $x_3(t)x_4(t)y_3(t)y_4(t) \to 0$ as $t \to \infty$.

The interpretation of Theorem 3 is that, in at least one of the populations, the frequency of at least one of the non-coalitionist types arbitrarily approaches zero, risking dying out. On the other hand, as we will see from the following example, Theorems 2 and 3 combined do not imply the coalitionist face is asymptotically stable (or even attracting) for the polymorphic model. This contrasts with Theorem 1 which showed the coalitionist vertex is asymptotically stable for the monomorphic model.

EXAMPLE 1. Let

$$\mathbf{A} := \begin{pmatrix} 10 & 1 \\ 2 & 4 \end{pmatrix}, \qquad \mathbf{B} := \begin{pmatrix} 2 & 4 \\ 1 & 6 \end{pmatrix}.$$

This is a *Coordination Game* (as a non-cooperative bimatrix game) and we will show that the coalitionist face is not globally attractive. This game has one mixed Nash equilibrium, $(\mathbf{p}^*, \mathbf{q}^*) = \left(\left(\frac{2}{3}, \frac{1}{3}\right), \left(\frac{3}{11}, \frac{8}{11}\right)\right)$ as well as two pure Nash equilibria, ((1,0),(1,0)) and ((0,1),(0,1)) . Since the matrix $\mathbf{A} + \mathbf{B}^{\mathrm{T}}$ has its strictly maximal entry

$$a_{11} + b_{11} = 12 \quad \text{and} \quad (\mathbf{p}^*\mathbf{A}\mathbf{q}^*, \mathbf{q}^*\mathbf{B}\mathbf{p}^*) = \left(\frac{38}{11}, \frac{8}{3}\right),$$

$(u, v) = (6.5, 5.5)$ is a strict NE-imputation. Furthermore, since maximal column sums of $\mathbf{A}$ and $\mathbf{B}$ are 12 and 10, respectively, this imputation satisfies the inequalities in (4.2) of Theorem 2. Thus the interior of the non-coalitionist face is a repellor.

For any $(\mathbf{x}, \mathbf{y})$ in the interior of $S_4 \times S_4$ near ((0,0,1,0), (1,0,0,0)), we have $(\bar{\mathbf{A}}y)_3 > (\bar{\mathbf{A}}\mathbf{y})_i$ for $i = 1, 2, 4$ and $(\overline{\mathbf{B}}\mathbf{x})_1 > (\overline{\mathbf{B}}\mathbf{x})_j$ for $j = 2, 3, 4$. Thus x_3 and y_1 are monotonically increasing in a neighbourhood of ((0,0,1,0),(1,0,0,0)) and so this rest point of (4.1) is neutrally stable. In fact, every interior trajectory of (4.1) initially sufficiently close to ((0,0,1,0),(1,0,0,0)) converges to a point on the edge E joining ((0,0,1,0),(1,0,0,0)) to

((0,0,1,0),(0,0,1,0)). Moreover, any initial interior point near the coalitionist pure strategy ((1,0,0,0),(1,0,0,0)) also evolves to a point on E near ((0,0,1,0),(1,0,0,0)). Thus the coalitionist face is not even locally attractive for this example.

The reason the coalitionist face is not asymptotically stable in Example 1 is that coalitionist populations, whose mean strategies are not near the interior "NE" in the coalitionist face when playing in a non-cooperative game (e.g. when they are near the pure coalitionist strategy ((1,0,0,0),(1,0,0,0))), evolve away from the coalitionist face. The following result shows this cannot happen when the coalitionist mean strategies are near the interior "NE" in the coalitionist face. Theorem 4 is then the counterpart of Theorem 1 for the monomorphic model in that in Section 3 the coalitionists were forced to play the interior NE strategy in their non-cooperative encounters.

Finally we notice that if we take $u = 5$ and $v = 7$ in the above example, then conditions of Theorem 4 hold but those of Theorems 2 and 3 do not.

THEOREM 4. *If* (u, v) *is a strict NE-imputation then each point of C near enough the point* $(\hat{\mathbf{x}}, \hat{\mathbf{y}}) := ((p_1^*, p_2^*, 0, 0), (q_1^*, q_2^* 0, 0))$ *is locally attractive for* $\text{int}(S_4 \times S_4)$.

Proof. Clearly we have

$$\hat{\mathbf{x}} = \left(\frac{b_{22} - b_{12}}{b_{11} - b_{12} - b_{21} + b_{22}}, \frac{b_{11} - b_{21}}{b_{11} - b_{12} - b_{21} + b_{22}}, 0, 0\right),$$

$$\hat{\mathbf{y}} = \left(\frac{a_{11} - a_{21}}{a_{11} - a_{12} - a_{21} + a_{22}}, \frac{a_{22} - a_{12}}{a_{11} - a_{12} - a_{21} + a_{22}}, 0, 0\right).$$

Consider the function $L^* : \text{int}(S_4 \times S_4) \to \mathbf{R}$

$$L^*(\mathbf{x}, \mathbf{y}) := \hat{x}_1 \ln x_1 + \hat{x}_2 \ln x_2 - \hat{x}_1 \ln x_3 - \hat{x}_2 \ln x_4 + \hat{y}_1 \ln y_1 + \\ + \hat{y}_2 \ln y_2 - \hat{y}_1 \ln y_3 - \hat{y}_2 \ln y_4.$$

Its derivative with respect to the polymorphic replicator dynamics (4.1) is

$$DL^*(\mathbf{x}, \mathbf{y}) = u(y_1 + y_2) - \mathbf{p}^* \mathbf{A}(y_1, y_2) + v(x_1 + x_2) - \mathbf{q}^* \mathbf{B}(x_1, x_2).$$

Now let us consider function L^* near the Nash-equilibrium lying in C. We claim that the derivative with respect to the dynamics (4.1) is positive. To see this, take $\boldsymbol{\varepsilon}, \boldsymbol{\delta} \in \mathbf{R}^4$ and define

$$\mathbf{x} = \hat{\mathbf{x}} + \boldsymbol{\delta} \text{ and } \mathbf{y} = \hat{\mathbf{y}} + \boldsymbol{\varepsilon}.$$

Using this notation we have

$$DL^*(\mathbf{x},\mathbf{y}) = u - \mathbf{p}^*\mathbf{A}\mathbf{q}^* + v - \mathbf{q}^*\mathbf{B}\mathbf{p}^* + u(\varepsilon_1 + \varepsilon_2) - \\ - \mathbf{p}^*\mathbf{A}(\delta_1, \delta_2) + v(\delta_1 + \delta_2) - \mathbf{q}^*\mathbf{B}(\varepsilon_1, \varepsilon_2).$$

Since (u, v) is a strict NE-imputation, for $|\boldsymbol{\varepsilon}|$ and $|\boldsymbol{\delta}|$ sufficiently small we have $DL^*(\mathbf{x},\mathbf{y}) > 0$. Thus the coalitionist face is attractive in the neighbourhood of the Nash equilibrium.

Observe the conclusion of Theorem 3 remains valid under the weaker assumption that the coalition game is NE-essential for a given imputation (i.e. $u + v > \mathbf{p}^*\mathbf{A}\mathbf{q}^* + \mathbf{q}^*\mathbf{B}\mathbf{p}^*$). □

Remark 2. The proof of the above theorem could also be carried out by linearizing the dynamics (4.1) about the interior NE within C.

5. DISCUSSION

The aim of this paper was to introduce the replicator dynamics approach into the theory of cooperative games. Cooperative game theory traditionally works with maximin solutions for given coalitions; whereas, the Nash equilibrium plays a central role for the replicator dynamics (e.g. every interior equilibrium of the replicator dynamics is a Nash equilibrium in the bimatrix game). For this reason, we developed our model and theorems based on the concept of Nash equilibrium rather than maximin solutions.

We investigated the propagation of coalitionist behaviour in terms of both monomorphic and polymorphic replicator dynamics for NE-essential games. We have shown that in the monomorphic model the coalitionist players eliminate the non-coalitionists. In the polymorphic model the coalitionists are able to propagate in the population but in general they cannot eliminate the non-coalitionist behaviour.

Finally, we note that a dynamic approach to coalition games may also be conceptually important in other biological contexts, such as those to describe the evolution of symbiotic and mutualistic interactions.

NOTE

1. In the theory of cooperative games, the standard definition of an essential game uses the maximin value instead of the present Nash equilibrium value.

REFERENCES

Akin, E. (1980), Domination or equilibrium, *Mathematical Biosciences* 50, 239–250.

Akin, E. and Hofbauer, J. (1982), Recurrence of the unfit, *Mathematical Biosciences* 60, 51–62.

Cressman, R. (1992), *The Stability Concept of Evolutionary Game Theory*. Berlin: Springer.

Hofbauer, J. and Sigmund, K. (1998), *Evolutionary Games and Population Dynamics*. Cambridge: Cambridge University Press.

Owen, G. (1968), *Game Theory*, Philadelphia: Saunders.

Owen, G. (1995), *Game Theory*, 3rd edn. San Diego: Academic Press.

Weibull, J.W. (1995), *Evolutionary Game Theory*. Cambridge, Mass.: The MIT Press.

Address for correspondence: Ross Cressman, Department of Mathematics, Wilfrid Laurier University, Waterloo, Ont., Canada N2L 3C5, (E-mail: rcressma@wlu.ca)

József Garay, Theoretical Biology and Ecological Modelling Research Group of the Hungarian Academy of Science and Department of Plant Taxonomy and Ecology, L. Eötvös University, Pázmány Péter sétány 1/C H-1117 Budapest, Hungary (E-mail: garayj@ludens.elte.hu)

Antonino Scarelli, Department of ecology and Sustainable Economic Development, University of Tuscia, Via S. Giovanni Decollato, 01100 Viterbo, Italy (E-mail: scarelli@unitus.it)

Zoltán Varga, Institute of Mathematics and Informatics, Szent István University, Páter K. u. 1. Godollo H-2103 Hungary
(E-mail: zvarga@ mszi.gau.hu)

EMIKO FUKUDA and SHIGEO MUTO

DYNAMIC COALITION FORMATION IN THE APEX GAME

ABSTRACT. This paper studies stability of coalition structures in the apex game by the use of Bloch's (1996) model of dynamic coalition formation. In our model, players' payoffs are given by the coalition values and a cost for proposal is newly introduced. We study stable coalition structures by the subgame perfect equilibrium and we show that in the apex game stable coalition structures depend on the cost for making a proposal and who the first proposer is. As opposed to the static analysis by Hart and Kurz (1983), it turns out that the apex may form a two-person coalition with a minor player.

1. INTRODUCTION

Owen (1977) defined the coalition value, that is, a generalized Shapley value (Shapley, 1953) with *a priori* coalition structure.

Hart and Kurz (1983) studied stability of coalition structures using the coalition value. Their models were presented as strategic form games, and the strong Nash equilibrium (Aumann, 1967) was used to study stable coalition structures. In particular, they studied stable coalition structures in the apex game, and showed that if the number of the players is greater than or equal to five, the coalition of all minor players is a unique stable coalition.

On the other hand, a dynamic process of coalition formation was investigated in Bloch (1996).

In this paper, we apply Bloch's dynamic coalition formation model to the apex game after slightly modifying the model. Players' payoffs are given by the coalition value and costs for making proposals are newly introduced. We use the subgame perfect equilibrium to study stable coalition structures.

Theory and Decision **56:** 153–163, 2004.

We show that in the apex game stability of coalition structures depends on the amount of the cost for proposal and who the first proposer is. As opposed to the static analysis by Hart and Kurz (1983), it turns out that the apex may form a two-person coalition with a minor player.

The paper is organized as follows. In Section 2, we provide definitions to be used in the following sections, and review the results of Hart and Kurz (1984). Section 3 is devoted to the description of a dynamic model of coalition formation and stability concept in this model. Stability of coalition structures in the apex game is also studied. In Section 4, we fully describe relations between the cost for a proposal and stable coalition structures by using a five-person example. We end the paper with some concluding remarks.

2. PRELIMINARIES

This section includes the definitions of a coalitional form game, a coalition structure and the coalition value that will be used in the following parts. After that, we review the static models of coalition formation proposed by Hart and Kurz (1983), and the stable coalition structures in their models.

2.1. *Coalition Value*

Let $N = \{1, 2, \ldots, n\}$ be the set of players. We define a coalitional form game with player set N by $v : 2^N \to \mathbf{R}$ satisfying $v(\phi) = 0$.

Let $\mathbf{B} = \{B_1, \ldots, B_m\}$ be a partition of player set N, that is, $\bigcup_{k=1}^{m} B_k = N$ and $B_k \cap B_l = \phi$ for all k, $l \in M = \{1, \ldots, m\}$, $k \neq l$. In the following, we call $\mathbf{B}$ a coalition structure.

For each coalitional form game v with coalition structure $\mathbf{B}$, Owen (1977) defines the coalition value for each player $i \in N$, i. e., a generalized Shapley value with a priori coalition structure.

Formally, for $B_j \in \mathbf{B}$ and $i \in B_j$,

$$\phi_i(v, \mathbf{B}) = \sum_{\substack{H \subseteq M \\ j \notin H}} \sum_{\substack{S \subseteq B_j \\ i \notin S}} \frac{h!(m-h-1)!s!(b_j-s-1)!}{m!b_j!} \times [v(Q \cup S \cup i) - v(Q \cup S)],$$

where h, s and b_j are the cardinalities of H, S and B_j, and $Q = \cup_{k \in H} B_k$.

The coalition value $\phi_i(v, \mathbf{B})$ is interpreted as the utility of player i of participating in the game v, when players are organized according to $\mathbf{B}$. The coalition value is derived from a system of axioms. See Owen (1977).

2.2. *Static Model*

A strategic form game model of coalition formation is presented by Hart and Kurz (1983). In their model, for a given coalitional form game v, each player proposes a coalition that he wants to form. Payoffs to players in each coalition structure are determined by the coalition value. The coalition structure is said to be stable if there is no group of players such that they are all better off by forming a new coalition.

Depending on the reaction of other players, two notions of stability are considered. In the γ-model, it is assumed that a coalition breaks up into singletons when at least one of its members leave. In the δ-model, remaining players continue to form a coalition.

Formally, for a coalitional form game v with the set of players $N = \{1, 2, \ldots, n\}$, the games of coalition formation, $\Gamma(v, N)$ and $\Delta(v, N)$, are defined as follows.

DEFINITION 1 (γ-model). The game $\Gamma(v, N)$ consists of the following:

(1) The set of players is N.
(2) For each $i \in N$, the set of strategies of player i is $\Sigma^i = \{S \subset N | i \in S\}$.
(3) For each n-tuple of strategies $\sigma^i = (S^1, S^2, \ldots, S^n) \in \prod_{i=1}^n \Sigma^i$ and each $i \in N$, the payoff to i is $\phi_i(v, \mathbf{B}_\sigma^{(\gamma)})$, where

$$T^i_\sigma = \begin{cases} S^i & \text{if } S^j = S^i \quad \text{for all } j \in S^i, \\ \{i\} & \text{otherwise} \end{cases}$$

and $\mathbf{B}^{(\gamma)}_\sigma = \{T^i_\sigma | i \in N\}$.

In the model γ, to form a coalition, all of its members must choose the coalition; and thus when a member (or members) deviates, the coalition breaks up.

DEFINITION 2 (δ-model). The game $\Delta(v, N)$ consists of (1), (2) and

(4) For each n-tuple of strategies $\sigma^i = (S^1, S^2, \ldots, S^n) \in \prod_{i=1}^n \Sigma^i$ and each $i \in N$, the payoff to i is $\phi_i(v, \mathbf{B}^{(\delta)}_\sigma)$, where $\mathbf{B}^{(\delta)}_\sigma = \{T \subset N | i, j \in T \text{ if and only if } S^i = S^j\}$.

In the model δ, any player who proposes the same coalition, forms a coalition. Thus in general, the coalition formed is not the one proposed by its members. Even if a member (or members) deviates, remaining players continue to form a coalition.

For a coalition structure $\mathbf{B}$ and a player $i \in N$, let $S^i_\mathbf{B}$ be the element of $\mathbf{B}$ to which i belongs and put $\sigma_\mathbf{B} = (S^i_\mathbf{B})_{i \in N}$. Here, we note that $S^i_\mathbf{B}$ is defined uniquely.

Hart and Kurz (1984) characterize the stable coalition structure by the notion of strong Nash equilibrium.

DEFINITION 3. The coalition structure $\mathbf{B}$ is γ-stable (respectively, δ-stable) in the game v with N if $\sigma_\mathbf{B}$ is a strong Nash equilibrium in $\Gamma(v, N)$ (respectively, $\Delta(v, N)$); i.e., if there exists no non-empty $T \subset N$ and no $\hat{\sigma}^i \in \Sigma^i$ for all $i \in T$, such that $\phi_i(v, \hat{\mathbf{B}}) > \phi_i(v, \mathbf{B})$ for all $i \in T$, where $\hat{\mathbf{B}}$ is produced by $((\hat{\sigma}^i)_{i \in T}, (\sigma^j_\mathbf{B})_{j \in N \setminus T})$ according to (3) (respectively, (4)).

2.3. *Statically Stable Coalition Structures in the Apex Game*

Hart and Kurz (1984) studied stable coalition structures (using their static models) in the apex game. The apex game is a special class of a voting game.

The n-player apex game consists of one "major" player and n-1 "minor" players. The major player is called the apex. This game has two types of winning coalitions; the one consists of the major player and at least one minor player, and the other consists of all minor players.

Formally, the apex game v is given by

$$v(S) = \begin{cases} 1 & \text{if } 1 \in S \text{ and } S\backslash\{1\} \neq \phi, \text{ or } S = N\backslash\{1\}, \\ 0 & \text{otherwise,} \end{cases}$$

where the set of players $N = \{1, 2, \ldots, n\}$ in which 1 is the apex.

In order to shorten notation, we sometimes denote coalition structures in a compact way: $\{\{1,2\},\{3\}\}$ is denoted by $[1\ 2 \mid 3]$, and so on.

In Hart and Kurz (1984), they obtained the following result:

PROPOSITION 1. *Let v be the n-player apex game. For $n \geq 5$, the unique γ-stable coalition structure is $[1 \mid 2\ 3 \ldots n]$, and there is no δ-stable coalition structures. For $n = 4$, $[1 \mid 2\ 3\ 4]$ is γ-stable but not δ-stable, and $[1\ 2 \mid 3 \mid 4]$, $[1\ 3 \mid 2 \mid 4]$ and $[1\ 4 \mid 2 \mid 3]$ are both γ-stable and δ-stable.*

3. DYNAMIC ANALYSIS

3.1. *Dynamic Model*

We now present an extensive form game model of coalition formation. This model is a special case of the model proposed by Bloch (1996).

The extensive form game of coalition formation with player set N proceeds as follows. The first proposer i is determined randomly and he starts the game by proposing coalition S such that $i \in S$. Here, whenever $|S| > 1$ the proposer must pay a cost for the proposal.

Each player in $S\backslash\{1\}$ responds to the proposal in an order determined randomly. If one of the players, say j, rejects the proposal, he must make a counteroffer and propose coalition S' such that $j \in S'$.

If all members accept, then coalition S is formed and all members of S withdraw from the game. Then the first proposer in

$N \backslash S$ is chosen randomly, and he starts to make a proposal. If the game continues infinitely, then all players receive a payoff of zero.

Note that, in this procedure, once a coalition has been formed the game is only played among the remaining players.

Figure 1 illustrates the extensive form game with three players.

The formal definition of the game $\Lambda(v, N)$ is as follows:

DEFINITION 4 (Dynamic model). The sequential coalition formation game $\Lambda(v, N)$ is a tuple $\langle N, H, P, c, (u_i)_{i \in N} \rangle$, in which

- N is the set of players.
- H is the set of histories, and Z denotes the set of terminal histories. At any point in the game $\Lambda(v, N)$, a history h determines:
  - a set $K(h)$ of players who have already formed coalitions,
  - a coalition structure $\mathbf{B}_h$ formed by the players in $K(h)$,
  - an ongoing proposal $T(h)$.
- $P : H \backslash Z \rightarrow N$ is the player function that assigns to each non-terminal history a member of N in the following manner: if $T(h) = \phi$ then $P(h) \in N \backslash K(h)$, otherwise $P(h) \in T(h)$.

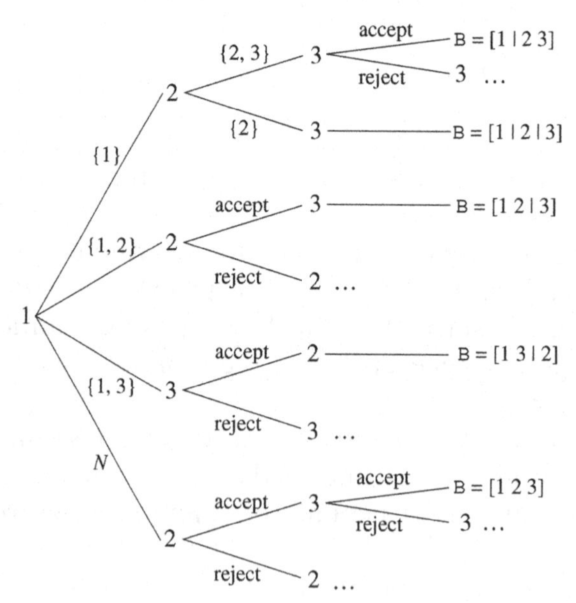

Figure 1. Dynamic model.

A strategy σ_i for player i is a mapping from H_i to his set of actions, namely

$$\begin{cases} \sigma_i(h) \in \{T \subset N \backslash K(h) | i \in N\} & T(h) = \phi, \\ \sigma_i(h) \in \{ \text{ accept, reject } \} & T(h) \neq \phi. \end{cases}$$

- c is a non-negative real number which stands for the cost for making an offer.
- u_i is the payoff function of i, given by, for each $\sigma = (\sigma_i)_{i \in N}$,

  $$u_i(\sigma) = \delta^{t(\sigma)} \phi_i(v, \mathbf{B}(\sigma)) - c_i(\sigma),$$

  where $\delta \in (0, 1)$ is a discount factor and $\mathrm{t}(\sigma)$ is the length of the terminal history which is determined by σ, and c_i (σ) is the unit cost c times the number of proposals player i makes if σ is played.

Then we define the stability concept of this dynamic model by the subgame perfect equilibrium.

DEFINITION 5. The coalition structure $\mathbf{B}$ is dynamically stable if it is produced by σ and σ is a subgame perfect equilibrium in $\Lambda(v, N)$, that is, for all $i \in N$, $h \in H_i$ and for all $\sigma_i'(h)$, $u_i(\sigma) \geq u_i(\sigma | \sigma_i'(h))$ holds, where $\sigma | \sigma_i'(h)$ means the strategy profile in which player i uses $\sigma_i'(h)$ at history h instead of $\sigma_i(h)$.

3.2. *Dynamically Stable Coalition Structures in the Apex Game*

We study stable coalition structures in the apex game in the dynamic model that we defined above. The main result is the following:

THEOREM 1. *Let v be the n-player apex game, and let δ be close enough to 1 and c be small enough. If the apex is the first proposer then $[1\ j | k | \ldots | n]$ is the unique stable coalition structure. If a minor player is the first proposer then $[1\ j | k | \ldots | n]$ and $[1 | 2 \ldots n]$ are stable.*

Proof. Let $\mathbf{B} = \{B_1, \ldots, B_m\}$ be a coalition structure attained by a subgame perfect equilibrium. Without loss of generality, we assume that $1 \in B_1$ and $k = |B_1|$. Two cases are considered; that is, $k \geq 2$ and $k = 1$.

Case 1: $k \geq 2$. We have

$$\phi_i(v, \mathbf{B}) = \frac{2}{mk(k-1)},$$

for all $i \in B_1 \backslash \{1\}$, and $\phi_i(v, \mathbf{B}) = 0$, for all $i \notin B_1$. See Hart and Kurz (1984) for details. In this case, player $i \notin B_1$ has no incentive to forming a coalition and to paying a cost. We obtain $|B_j| = 1$ for all $B_j \neq B_1$ in the equilibrium outcome. Therefore, we must have $m = n - k + 1$.

Case 2: $k = 1$. We see

$$\phi_i(v, \mathbf{B}) = \frac{2}{m(m-1)|B_j|}, \quad i \in B_j \neq B_1,$$

for all members i of B_j with $j \neq 1$.

Assume that $m \geq 3$. Take $j \neq 1$ such that for all $i \in B_j$, $\phi_i(v, \mathbf{B}) < \phi_l(v, \mathbf{B})$ for all $l \in N \backslash (B_j \cup B_l)$. For $i \in B_j$ if $\{i, 1\}$ is formed he can receive $1/m$. On the other hand, in both cases of $k \geq 2$ and $k = 1$,

$$\phi_1(v, \mathbf{B}) = 1 - \frac{2}{mk}.$$

If $\{1, i\}$ is formed player 1 can receive a higher payoff. Therefore there exists a history h such that $P(h) = 1$ or i and the action at h is $\{1, i\}$ which increases their payoffs in equilibrium.

Hence $m = 2$ must hold in any subgame equilibrium outcome.

We will now consider separately two cases depending on who the first proposer is.

Let the first proposer be a minor player and denote this player by i. If the first proposal is a k-person coalition including the apex, he will have a payoff of

$$\frac{2}{(n-k+1)k(k-1)}.$$

On the other hand, if the first proposal is a coalition which consists of all minor players, he will have $1/(n-1)$. Simple calculation shows that

$$\frac{2}{(n-k+1)k(k-1)} < \frac{1}{n-1} \quad \text{if } 2 < k < n$$

and

$$\frac{2}{(n-k+1)k(k-1)} = \frac{1}{n-1} \quad \text{if } k = 2.$$

Therefore for each minor player i, both B_1 with $k = 2$ and $\{2, 3, \ldots, n\}$ are the best as the first proposal, which proves the second case of the theorem.

Let the first proposer be the apex. The apex can obtain a positive amount whenever he proposes a coalition B_1 such that $k \geq 2$. However, minor players accept only the coalition with $k = 2$; otherwise they reject the proposal of the apex and propose $\{2, 3, \ldots, n\}$. Hence the apex proposes B_1 such that $k = 2$. The proof is complete. □

REMARK. If the cost $c > \frac{1}{n-1}$ then every minor player withdraws from the game by proposing the singleton coalition in his turn.

Comparing with Proposition 1, two features of the dynamical analysis can be pointed out. First, dynamically stable coalition structures depend on who the first proposer is. This implies that it is very important for both the apex and minor players to get the initiative of the bargaining. Secondly, when c is small enough, even if the apex gets the initiative he makes a compromise as to his payoff and forms a coalition with one minor player; otherwise, he has payoff of zero.

4. EXAMPLE

In this section, we describe stable coalition structures in the five-person apex game and make clear the relations between the cost for proposal and stable coalition structures.

Let $N = \{A, B, C, D, E\}$ and A be the apex. Let v be the apex game with player set N. We can also describe this game [4; 3, 1, 1, 1, 1]. Coalition values for each coalition structure are given in Table I.

TABLE I
Coalition values

Coalition structure	Coalition value
[A B C D E] [A | B | C | D | E]	(3/5, 1/10, 1/10, 1/10, 1/10)
[A | B C D E]	(0, 1/4, 1/4, 1/4, 1/4)
[A B | C | D | E]	(3/4, 1/4, 0, 0, 0)
[A B C | D | E]	(7/9, 1/9, 1/9, 0, 0)
[A B C D | E]	(3/4, 1/12, 1/12, 1/12, 0)

Let us apply our dynamic model and find the stable coalition structures for each cost range. Firstly, let $c \leq 5/36$. If the apex is the first proposer then the coalition structure [A B | C | D | E] is essentially the unique stable coalition structure. If one of the minor players is the first proposer then [A B | C | D | E] and [A | B C D E] are stable.

Let $5/36 < c \leq 1/4$. If a minor player is the first proposer [A B | C | D | E] and [A | B C D E] are stable. And if the apex is the first proposer then [A B C | D | E] is essentially the unique stable coalition structure, because the minor player cannot make any counter offer.

Let $c > 1/4$. Here no minor player proposes to form a coalition with other players. So, if the apex is the first, the second or the third proposer [A B C | D | E] is the unique stable coalition structure, and if he is the fourth proposer then [A B| C | D | E] is stable, and if he is the final proposer then [A | B | C | D | E] is stable.

5. CONCLUDING REMARKS

In Hart and Kurz (1984) they studied stability of coalition structures in the apex game by using static models of coalition formation.

In this paper a dynamic model of coalition formation by Bloch (1996) is employed. We have given players' payoffs by Owen's coalition value and introduced a cost for each proposal in this model. And we have used the subgame perfect equilibrium to study stable coalition structures.

It was observed that in the apex game stability of coalition structures depends on the amount of the cost for proposal and who the first proposer is. Especially, we found that it is very important for minor players and also for the apex to get the initiative in the bargaining.

Finally we fully describe stable coalition structures in the five-person apex game.

REFERENCES

Aumann, R.J. (1967), A survey of cooperative games without side payments, in M. Shubik (ed), *Essays in Mathematical Economics*. Princeton: Princeton University Press, 3–27.

Bloch, F. (1996), Sequential formation of coalition with fixed payoff division, *Games and Economic Behavior* 14, 90–123.

Hart, S. and Kurz, M. (1983), Endogenous formation of coalitions, *Econometria* 51, 1047–1064.

Hart, S. and Kurz, M. (1984), Stable coalition structures, in M.J. Holler (ed.), *Coalitions and Collective Action*. Wurzburg: Physica-Verlag, 236–258.

Owen, G. (1977), Values of games with a priori unions, in R. Hein and O. Moeschlin (eds), *Essays in Mathematical Economics and Game Theory*. New York: Springer-Verlag, 76–88.

Shapley, L.S. (1953), A value for *n*-person games, in H.W. Kuhn and A.W. Tucker (eds), *Contributions to the Theory of Games*. Princeton: Princeton University Press, 307–317.

Address for correspondence: Emiko Fukuda, Departement of Value and Decision Science, Graduate School of Decision Science and Technology, Tokyo Institute of Technology, 2-12-1 Ookayama, Meguro-ku, Tokyo 152-8552, Japan. (E-mail: emiko@valdes.titech.ac.jp).

Shigeo Muto, Departement of Value and Decision Science, Graduate School of Decision Science and Technology, Tokyo Institute of Technology, 2-12-1 Ookayama, Meguro-Ku, Tokyo 152-8552, Japan.
(E-mail: muto@valdes.titech.ac.jp).

YAROSLAVNA PANKRATOVA and SVETLANA TARASHNINA

HOW MANY PEOPLE CAN BE CONTROLLED IN A GROUP PURSUIT GAME

ABSTRACT. In this paper we study a time-optimal model of pursuit. The game is supposed to be a nonzero-sum simple pursuit game between a pursuer and m evaders acting independently of each other. Here we assume that the evaders are discriminated and dictated the extremely disadvantageous behaviour by the pursuer who has an element of punishment at his disposal. The aim of this paper is to find an equilibrium situation and to define the maximum number of participants of the game that can be kept in submission. For every evader we construct a realizability area of punishment strategy and investigate the question of its existence depending on various initial positions of the players P and E_i, $i = 1, \dots, m$.

1. INTRODUCTION

Differential games represent conflict situations with an infinite set of alternatives. They can be described with the help of differential equations. The main reason for using differential games used to be the necessity to solve military problems. However, recently differential games are also applied to a wider range of problems.

When only one pursuer and one evader participate in the process of pursuit, we deal with a classical zero-sum differential pursuit game. If there are more than two players involved in a conflict situation, then modelling such a problem as a nonzero-sum game seems to be more natural and adequate. Although even in this case most authors only considered zero-sum models, dividing all the players into two groups with opposite interests and taking each group as a separate player. For the first time, zero-sum differential pursuit games were described with many examples in Isaacs (1965).

Theory and Decision **56:** 165–181, 2004.

However, players' goals are not always strictly opposed. Presently differential games are not only used for solving military problems, but also for solving economical and psychological problems. In other words, we will study those conflict situations which involve more than two players that are not automatically inclined to destruct each other. These are the so called nonzero-sum pursuit games and they are typically used to model less aggressive behaviour. Up to now this sort of models has not been paid much attention. Nevertheless, for example in case there are many evaders in a pursuit game, each of which wanting to maximize his own capture time, the assumption that all the evaders will be forming one big coalition for a long time seems to be unrealistic. In this case the construction of a zero-sum model seems to be a bit artificial and the non-zero model seems to be the more natural choice.

In this paper we examine a nonzero-sum differential game of simple pursuit between a pursuer and m evaders acting independently of each other. We suppose that the pursuer has a strategy at his disposal that allows him to dictate to the evaders what motion to choose. The pursuer can impose certain behaviour on the evaders that can be adverse to them by threatening to change the order of pursuit. So, in this paper we study the conflicts with many participants, one of which is stronger than the others and who can affect the behaviour of the others. Such problems can be interpreted in the following way: the so called leader (the pursuer), possesses a certain power, and has some instrument of punishment at his disposal with which he can force a certain number of people (the evaders) with less opportunities to do precisely what he likes. The aim of this paper is to find the equilibrium in this situation and to define the maximum number of participants of the game that can be kept in submission.

Here we assume that the evaders are dictated the extremely disadvantageous behaviours. In real life we face this kind of situations quite often. As an example one can think of the so called totalitarian state whose citizens have two options: either to obey or to be punished. The principal-agent relationship in a working group can also be considered as an illustrative example for modelling a psychological aspect of this problem.

2. THE MODEL

The game under study is a time-optimal model of pursuit in which $(m+1)$ players – one pursuer P and m evaders $E_1, \dots, E_m$ – move on a plane with constant velocities. The players start their motion at the position $z_p^0 = (x_p^0, y_p^0)$, $z_i^0 = (x_i^0, y_i^0)$ and have the possibility of making decisions continuously in time, i.e. at each time instant they may choose the direction of their motion (velocity-vectors). We describe the case of perfect information. This means that each player at each time instant $t \geq 0$ knows the time t and his own as well as all other player' s positions. Additionally, we assume that the pursuer knows the vector-speeds chosen by the evaders at that time moment. So, in that sense the pursuer has the advantage.

Denote by $P^t = z_P^t = (x_P^t, y_P^t)$ and by $E_i^t = z_i^t = (x_i^t, y_i^t)$ the current positions of pursuer P and evader E_i at time instant $t > 0$, respectively.

The players use a simple motion, described by the following system of differential equations

$$\begin{aligned} &\text{for } P: \dot{z} = u_p, \quad u_p \in U_p; \\ &\text{for } E_i: \dot{z}_i = u_i, \quad u_i \in U_i, \end{aligned} \tag{1}$$

with initial conditions

$$z_p(0) = z_p^0, \quad z_i(0) = z_i^0, \quad i = 1, \dots, m, \tag{2}$$

where $z_i, z_p \in \mathbf{R}^2$, and the sets of control variables U_p, U_i have the following forms

$$U_p = \left\{ u_p = (u_p^1, u_p^2) : (u_p^1)^2 + (u_p^2)^2 \leq \alpha^2 \right\},$$

$$U_i = \left\{ u_i = (u_i^1, u_i^2) : (u_i^1)^2 + (u_i^2)^2 \leq \beta_i^2 \right\}, \quad i = 1, \dots, m.$$

Here β_i, α are constants, and it is common knowledge that $\alpha > \max_{i=1,\dots,m} \beta_i$. The velocity-vectors $u_p \in U_p$ and $u_i \in U_i$ are the control variables of the players P and E_i, respectively.

It is obvious that, being faster in speed, the pursuer can always ensure consecutive capture of all evaders within finite time. In the present paper, by capture we mean coincidence of players' positions.

Now we need to explain how players select their control variables throughout the game according to the incoming information. In other words, we need to define the notion of a strategy in this game. It is well-known that the strategy must describe behaviour of a player in all information states in which he may find himself during the game. Thus, we regard a strategy of a player as a function of time and state variables with values in the set of controls. More specifically, a strategy of player E_i is a function $u_i(t, z_p^t, z_1^t, \ldots, z_i^t, \ldots, z_m^t)$ satisfying system (1) with initial conditions (2). All these strategies belong to the strategy set of E_i. Denote it by $\mathcal{U}_i = \{u_i(\cdot)\}$.

The pursuer's strategy is a function of time, players' positions and velocity-vectors of the evaders $u_p(t, z_p^t, z_1^t, \ldots, z_m^t, u_1, \ldots, u_m)$.

However, the pursuer's problem is more complicated than that. What he must do first is to define the order of pursuit, which is a function of initial positions. Obviously, these two components of the pursuer's strategy have to be defined before the pursuer can play this game. The solution of this game was obtained in Petrosjan and Tomskii (1983).

We extend this problem and construct a specific equilibrium. For this purpose we introduce the third component, a "punishment" component, into the pursuer's strategy. This component plays a major role in forming the outcome of the game. Consider the following definition.

DEFINITION 1. Let $\pi(z^0, u_1(\cdot), \ldots, u_m(\cdot))$ be a pursuit order chosen by the pursuer at the initial instant $t = 0$ for some fixed strategy profile of the evaders; let $u_p(t, z)$, $t \geq 0$ be a direct pursuit strategy and let $p = p(t, u_1(t), \ldots, u_m(t))$ be an element of punishment that consists in changing the pursuit order. Then we say that the triple $u^\pi = \langle \pi, u_p, p \rangle$ is a strategy of pursuer P and refer to it as the punishment strategy of P.

It can easily be seen that a strategy of pursuer P depends on time as well as the positions (including initial) and velocity-vectors of the evaders. Let Π be the set of all possible orders. Then the strategy set of P can be written as $\mathcal{U}_p = \{u^\pi\}_{\pi \in \Pi}$.

Note that at each time instant $t > 0$ the element of punishment coincides with π if all evaders follow the originally chosen strategy profile, i.e. $p = p(t, u_1(t), \ldots, u_m(t)) = \pi$, and it differs from π if any of the evaders changes the original strategy profile, i.e. $p = p(t, \bar{u}_1(t), \ldots, \bar{u}_m(t)) = \bar{\pi}$, where $\pi, \bar{\pi} \in \Pi$.

We assume that the pursuer aims to terminate the game as soon as possible. Under termination we mean capture of all evaders. In turn each evader wants to avoid his own capture as long as possible and does not care about the other evaders.

Denoting by K_p the payoff function of P, and by K_i the payoff function of evader E_i, we have

$$K_i(u^\pi, u_1, \ldots, u_i, \ldots, u_m) = \sum_{k \leq i} T_k^\pi,$$

where $T_k^\pi = \frac{|N^{k-1} E_k^{T_{k-1}}|}{\alpha - \beta_k}$ is the time spent by the pursuer of evader E_k, according to the pursuit order $\pi \in \Pi$, and $N^0 = P^0$, $k = 1, \ldots, m$. Here i is a number of the evader in the pursuit order $\pi = \{E_1, \ldots, E_m\}$. Now and then by $|AB|$ we denote the Euclid distance between points A and B in R^2.

The payoff of P is defined as the negative value of the payoff of evader E_i that is caught last. Thus, player P aspires to minimize the total pursuit time, while each evader wants to be caught as late as possible. Thus,

$$K_p(u^\pi, u_1, \ldots, u_i, \ldots, u_m) = -T^\pi,$$

where $T^\pi = \sum_{k=1}^m T_k^\pi$ is the total pursuit time, and π the chosen pursuit order.

In fact, we have defined a whole family of games, each game depending on a choice of the initial positions of the players. Let us fix the players' initial positions and consider the game $\Gamma(z_p^0, z_1^0, \ldots, z_m^0)$.

3. NASH EQUILIBRIUM IN THE GAME $\Gamma(z_p^0, z_1^0, \ldots, z_m^0)$

The considered game $\Gamma(z_p^0, z_1^0, \ldots, z_m^0)$ is a nonzero-sum game. The theory of nonzero-sum games does not have a unified approach to solution concepts for such games. There are actually a number of such concepts, each of which is based on some additional assumptions for players' behaviour and structure of the game. One of them is the well-known concept of Nash equilibrium. However, in some games this concept gives us simply absurd solutions. In our game this is the case.

It was proved in Tarashnina (1998) that in the game $\Gamma(z_p^0, z_1^0, \ldots, z_m^0)$ there exists a whole family of Nash equilibria that includes some which are extremely adverse to the evaders' interests, and some which are favourable for them, as well as all intermediate equilibria. The main problem there was not only to construct a Nash equilibrium, say $(u^{\pi*}(\cdot), u_1^*(\cdot), \ldots, u_m^*(\cdot))$, but also to prove that it is a Nash equilibrium for each current subgame $\Gamma(z_p^t, z_1^t, \ldots, z_m^t)$ starting at the time instant $t > 0$ along the optimal path. It was done for the whole set of Nash equilibria.

In the present paper we construct the extremely odd Nash equilibrium that is the most disadvantageous for the evaders among all the equilibria, find the conditions that support it and then, using simulation, try to answer the question for how many evaders this phenomena would take place.

The game is played as follows: at the initial moment of time the pursuer dictates to the evader a certain behaviour and chooses some pursuit order. After that he pursues the evaders according to the chosen order. As soon as any of the evaders deviates, the pursuer punishes him by changing the pursuit order, and starts chasing the deviant player. In other words, the pursuer fixes some pursuit order and calculates the total pursuit time taking into account that the evaders use the prescribed behaviour. Thus, he considers the $m!$ various pursuit orders and chooses the one that gives him the shortest total pursuit time. After that, P consequently pursues the evaders according to the chosen order and changes it as soon as any of the evader chooses a direction of motion different from the one dictated by the pursuer.

In fact, the situation is the following: the pursuer wants the evader to follow some specific behaviour. He fixes some strategy profile, and then the evaders are told to follow it. If someone does not, then he is punished by being chased first. It turns out that even a delusion of P can be a NE-outcome under some conditions. The interesting fact is that the punishment strategy can make any strategy profile of the evaders into a

Nash equilibrium under some conditions. The main point is to find the conditions under which all this makes sense.

So, let us consider the following strategy profile $(u_1(\cdot), u_2(\cdot), \ldots, u_m(\cdot))$, according to which each evader should go to the capture point of a currently pursued evader (in other words, to the point where the pursuer is coming soon). We want to show that this can be a NE-outcome.

Now let us describe the following types of behaviours to the evaders:

- behaviour $[u_i^{j'}]$ prescribes to move along the straight line connecting his own and the pursuer's current positions in the direction from P (to the current capture point $N^{j'}$);
- behaviour $[u_i^{j}]$ prescribes to move along the straight line to the capture point of the currently pursued evader $E_i^{j'}$, $j > j'$, namely, to the current capture point $N^{j'}$, where $N^{j'} = P^{T_{j'}}$, $j' \in \{1, \ldots, m\}$.

Let us denote the set of all evaders by $M = \{E_1, \ldots, E_m\}$ and the set of evaders that are not caught yet by $S = \{E_i^j\}_{j>j'}$. Here E_i^j-type of the evader is the jth in the line of pursuit evader among the ones not yet caught, and $E_i^{j'}$ is the evader who is currently pursued. We will refer to him as the current evader.

It is obvious that throughout the game at some moment $\bar{t}_i \geq 0$ each evader E_i changes its type from E_i into $E_i^{j'}$. So, the strategy $u_i(\cdot)$ of evader E_i can be described as a pair $u_i(\cdot) = \langle [u_i^j]_{0 \leq t < \bar{t}_i}, [u_i^{j'}]_{t \geq \bar{t}_i} \rangle$ of consequent behaviours. During the game evader E_i consequently uses both types of behaviours. Evader E_1 uses just type $[u_i^{j'}]$ as he becomes the current evader at once, at the very first moment. From now on, we will consider this strategy profile $(u_1(\cdot), u_2(\cdot), \ldots, u_m(\cdot))$ of the evaders.

4. INEQUALITIES PROVIDING THE EXISTENCE OF NASH EQUILIBRIA

Player P makes his choice of the pursuit order by taking into account all orders and choosing the one giving the smallest total pursuit time. Suppose that P has chosen the optimal pursuit order π^* and according to it starts to pursue the evaders. Without loss a generality we assume that he uses the strategy u^{π^*}, i.e. the optimal pursuit order is $\pi^* = \{E_1, \ldots, E_m\}$. Otherwise, we always can renumber the evaders according to $\pi^* = \{E_1, \ldots, E_m\}$. Hence, we have

$$T^{\pi^*} = \min_{\pi \in \Pi} T^{\pi} \tag{3}$$

Now we need to find the conditions under which the lifetime of an obedient evader (when he follows the recommendations of the pursuer) is more than the lifetime of a disobedient evader (when he deviates from the strategy dictated by the pursuer). For any E_i $(i = 2, \ldots, m)$ this is true if the following inequality holds

$$\frac{|N^{i-2}E_{i-1}^{T_{i-2}}|}{\alpha - \beta_{i-1}} + \frac{|N^{i-1}E_i^{T_{i-1}}|}{\alpha - \beta_i} > \frac{|N^{i-2}E_i^{T_{i-2}}|}{\alpha - \beta_i} \tag{4}$$

These two formulas provide existence of the given Nash equilibrium.

Now let us consider Figure 1: axis Ox is going along $P^0E_1^0$ and axis Oy is going down. Denote by φ^i the angle between Ox and the straight line N^iN^{i+1}. Introduce the following notations: $E_i^0 = (x_i^0, y_i^0)$, $N^i = (x_{N^i}, y_{N^i})$ for all $i \in \{1, \ldots, m\}$, and $E_i^{T_k} = (x_i^{T_k}, y_i^{T_k})$ for all $i > k$, $i \in \{1, \ldots, m\}$, $k \in \{1, \ldots, m\}$, $T_0 = 0$. The coordinate $E_i^{T_k}$ is a position of evader E_i at the moment T_k, where T_k $(k = 1, \ldots, m)$ are the capture moments of the previously caught evaders.

As a result we have m capture points $N^1, N^2, \ldots, N^m$ that correspond to the capture moments $T_1, T_2, \ldots, T_m$.

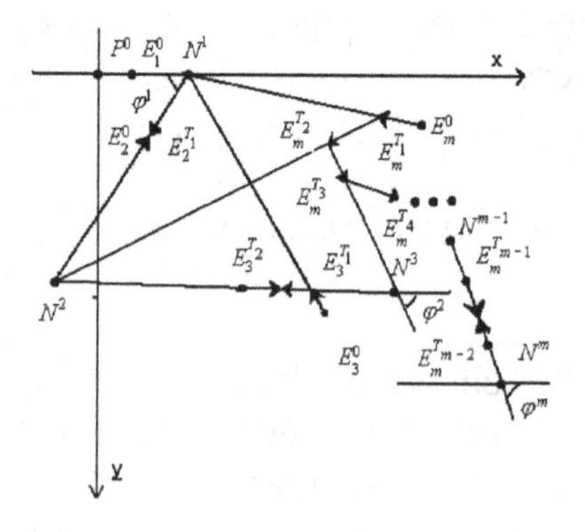

Figure 1.

In words, the equilibrium would look in the following way: at the moment $t = 0$ all evaders have to go to the first capture point N^1. Then the rest $(m - 1)$ evaders have to go to the second capture point N^2, and then to the next, and so on. On the last step we have just one evader left who has to go to the final capture point N^m.

So, in the form of coordinates formula (4) becomes

$$\frac{\alpha-\beta_i}{\alpha-\beta_{i-1}}\sqrt{\left(x_{N^{i-2}}-x_{i-1}^{T_{i-2}}\right)^2+\left(y_{N^{i-2}}-y_{i-1}^{T_{i-2}}\right)^2}+$$
$$\sqrt{\left(x_{N^{i-1}}-x_i^{T_{i-1}}\right)^2+\left(y_{N^{i-1}}-y_i^{T_{i-1}}\right)^2}>\sqrt{\left(x_{N^{i-2}}-x_i^{T_{i-2}}\right)^2+\left(y_{N^{i-2}}-y_i^{T_{i-2}}\right)^2}.$$

Now let us rewrite formula (3) in the coordinate form. The time spent by the pursuer chasing one particular evader E_i is

$$T_i^{\pi*}=\frac{|N^{i-1}E_i^{T_{i-1}^{\pi*}}|}{\alpha-\beta_i}=\frac{\sqrt{\left(x_{N^{i-1}}-x_i^{T_{i-1}}\right)^2+\left(y_{N^{i-1}}-y_i^{T_{i-1}}\right)^2}}{\alpha-\beta_i}.$$

Here, according to Figure 1, we have

$$x_{N^i}=x_{N^{i-1}}-\alpha T_i\cos\varphi^{i-1},\quad y_{N^i}=y_{N^{i-1}}-\alpha T_i\sin\varphi^{i-1},$$

where

$$\cos\varphi^{i-1}=\frac{x_{N^{i-1}}-x_i^{T_{i-1}}}{\sqrt{\left(x_{N^{i-1}}-x_i^{T_{i-1}}\right)^2+\left(y_{N^{i-1}}-y_i^{T_{i-1}}\right)^2}},$$

$$\sin\varphi^{i-1}=\frac{y_{N^{i-1}}-y_i^{T_{i-1}}}{\sqrt{\left(x_{N^{i-1}}-x_i^{T_{i-1}}\right)^2+\left(y_{N^{i-1}}-y_i^{T_{i-1}}\right)^2}}.$$

The coordinates of $E_i^{T_k}$ $(i = 1, \ldots, m, \quad k = 1, \ldots, i-1)$ are

$$x_i^{T_k}=x_i^{T_{k-1}}-\beta_i T_k\cos\varphi^i,\quad y_i^{T_k}=y_i^{T_{k-1}}-\beta_i T_k\sin\varphi^i,$$

where

$$\cos\varphi^i=\frac{x_{N^k}-x_i^{T_{k-1}}}{\sqrt{\left(x_{N^k}-x_i^{T_{k-1}}\right)^2+\left(y_{N^{k-1}}-y_i^{T_{k-1}}\right)^2}},$$

$$\sin \varphi^i = \frac{y_{N^k} - y_i^{T_{k-1}}}{\sqrt{\left(x_{N^k} - x_i^{T_{k-1}}\right)^2 + \left(y_{N^k} - y_i^{T_{k-1}}\right)^2}}.$$

So, we got recurrent formulas for calculating the existence conditions for the given Nash equilibrium

THEOREM 1. *Consider the game* $\Gamma(z_p^0, z_1^0, \ldots, z_m^0)$. *If conditions* (3) *and* (4) *are verified for all* $i = 1, \ldots, m$, *then there exists a Nash equilibrium. It is constructed as follows:*

1. *evader* E_i *uses the strategy* $u_i(\cdot)$ *that dictates to him*
   - *to choose type* $[u_i^{j'}]$ if $i = j'$, $j' \in \{1, \ldots, m\}$;
   - *to choose type* $[u_i^{j}]$ if $i = j$, $j > j'$, $j' \in \{1, \ldots, m\}$;
2. *pursuer* P, *according to the strategy* $u^{\pi^*}(\cdot)$, *minimizes the total pursuit time if each evader* E_i *adheres to the strategy* $u_i(\cdot)$ $(i = 1, \ldots, m)$, *and* P *changes the pursuit order as soon as any of the evaders* E_i^j $(j > j')$ *that are not yet caught deviates from the strategy* $u_i(\cdot)$ *dictated to him.*

It is clear that the evaders change their types at the capture moments T^i, $i = 1, \ldots, m$. For evader E_i this point comes at $\bar{t}_i = T^{i-1}$.

Theorem 1 gives us the conditions that support this odd Nash equilibrium. One of the points of the present paper is to describe geometrically a set of initial positions of the evaders for which the punishment strategy of P is realizable. In other words, we want to construct sets of initial positions for the evaders that support the Nash equilibrium $(u^{\pi^*}, u_1(\cdot), u_2(\cdot), \ldots, u_m(\cdot))$.

5. THE REALIZABILITY AREA OF THE PUNISHMENT STRATEGY

Further we need the following notations.

DEFINITION 2. The punishment strategy of pursuer P is called realizable with respect to evader E_i if conditions (3), (4) of Theorem 3 holds for this particular i.

DEFINITION 3. The punishment strategy of pursuer P is called completely realizable in the game $\Gamma(z_p^0, z_1^0, \dots, z_m^0)$ if it is realizable with respect to all evaders E_i, $i = 1, \dots, m$.

In other words, the punishment strategy is realizable whenever it forces some strategy profile of the evaders to be a Nash equilibrium.

First of all, we introduce the notion of "realizability area". For this purpose we associate with each evader E_i an area $\Omega_i \in R^2$ and refer to it as the realizability area. Let us fix the initial positions of all the other players. Area Ω_i is the set of all E_i's initial positions such that, when there, evader E_i has to adhere the strategy $u_i(\cdot)$ dictated by the pursuer.

We consider some examples and construct realizability areas of the punishment strategy because it seems to be interesting to know what form they would have.

EXPERIMENT 1. Let $\alpha = 100$, $\beta_1 = \beta_2 = 1/2$. Suppose also that P pursues evader E_1 first and then evader E_2. Let us fix the initial positions of the players E_1 and $P : P^0 = (0,0)$, $E_1^0 = (1,0)$ and define the set of positions $E_2^0(x,y)$ such that the situation $(u^{\pi^*}, u_1^*, u_2^*)$ is a NE-outcome. In other words, we construct the set of all positions $E_2^0(x,y)$ under which the punishment strategy of pursuer P is realizable with respect to evader E_2.

The total pursuit time with respect to the orders $\pi = \{1,2\}$ and $\pi' = \{2,1\}$ is calculated by the following formulas

$$T_\pi = \frac{|P^0 E_1^0|}{\alpha - \beta_1} + \frac{|N^1 E_2^{T_1}|}{\alpha - \beta_2} \quad \text{and} \quad T_{\pi'} = \frac{|P^0 E_2^0|}{\alpha - \beta_2} + \frac{|N^2 E_1^{T_1}|}{\alpha - \beta_1},$$

Now suppose that the optimal pursuit order is $\pi^* = \pi = \{1,2\}$. If it is not really so, we can renumber the evaders according to it. Hence, for our example condition (4) takes the following form

$$T_\pi < T_{\pi'} \tag{5}$$

Substituting the initial positions into inequalities (4) and (5), we have the following system of inequalities

$$\frac{198}{199}\left(1-\sqrt{x^2+y^2}\right) > \sqrt{\left(\frac{200}{199}x-1\right)^2+\left(\frac{200}{199}y\right)^2} -$$
$$-\sqrt{\left(\frac{200}{199}-x\right)^2+y^2}, \quad \frac{198}{199}+\sqrt{\left(\frac{200}{199}-x\right)^2+y^2} > \sqrt{x^2+y^2}.$$

Then, we obtain area Ω_2, with a shape as that of Figure 2 (left). If $E_2^0(x,y) \in \Omega_2$, then under these initial conditions the situation $(u^{\pi^*}, u_1^*, u_2^*)$ is a NE-outcome, and hence, pursuer P can dictate the described adverse behaviour to evader E_2. All this means that the punishment strategy of player P is realizable with respect to evader E_2 staying in area Ω_2. Then, we consider the game $\Gamma(z_p^0, z_1^0, z_2^0, z_3^0)$ and define the set of all positions $E_2^0(x,y)$ such that the situation $(u^{\pi^*}, u_1^*, u_2^*, u_3^*)$ is a NE-outcome. For this purpose we fix the point $E_2^0(x,y) \in \Omega_2$. Let the speed of evader E_3 be equal to $1/2$. So, we have $\alpha = 100$, $\beta_1 = \beta_2 = \beta_3 = 1/2$, $P^0 = (0,0)$, $E_1^0 = (1,0)$ and $E_2^0 = (10,10)$. Doing the computations we did before, we obtain a very complex system of inequalities. The analytical form of the system does not give us any clue about the form and properties of the considered area. Moreover, it is impossible even to say whether this area is empty or not.

Note that such a problem already occurs when there are just four players in the game ($n = 4$). It is obvious that in case $n > 4$ the system of inequalities becomes even more complicated (if

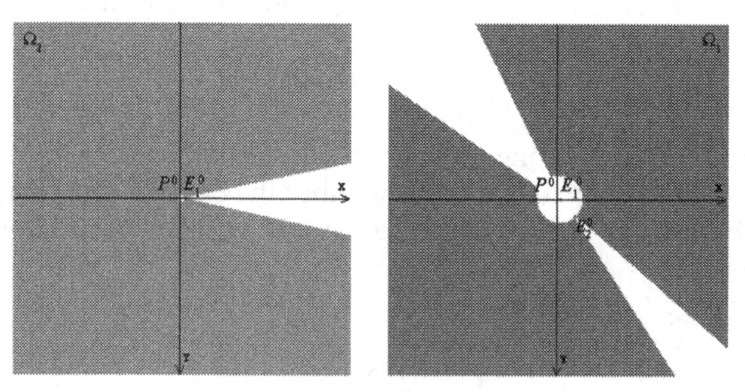

Figure 2.

$n = 5$, then the number of possible pursuit orders is equal to 24).

In order to handle this problem we create an executable computer program that solves the system of inequalities and construct the corresponding areas.

For $m = 3$ we have area Ω_3 of the initial positions for evader E_3. In other words, we define the set of all positions $E_3^0(x, y)$ such that the situation $(u^{\pi^*}, u_1^*, u_2^*, u_3^*)$ is a NE-outcome (Figure 2 (right)). Note that area Ω_3 is smaller than area Ω_2. This means that, as the number of evaders increases, their associated realizability areas Ω_i shrink.

Further we fix the point E_3^0 in Ω_3 and construct the next area. Let the speed of E_4 be also equal to $\beta_4 = 1/2$. Area Ω_4 is as in Figure 2 (right). Notice also that Ω_4 is narrower than Ω_3.

Let us continue doing this in the same way in order to find out the very first number i for which Ω_i becomes empty (beyond which further consideration of this problem is senseless).

We select from Ω_4 the point $E_4^0 = (-200, -200)$ and construct Ω_5 for evader E_5. As before, using the computer program we construct area Ω_5 (Figure 3 (right)). This area is narrower than those above constructed, Ω_2, Ω_3 and Ω_4.

Further, doing similar computations, we fix the point $E_6^0 = (500, 500) \in \Omega_6$ (Figure 4 (left)) and construct area Ω_7 (Figure 4 (right)). We can see that area Ω_7 of initial positions $E_7^0(x, y)$ is empty (Figure 4 (right)). In other words, the punishment strategy of the pursuer is not realizable with respect to

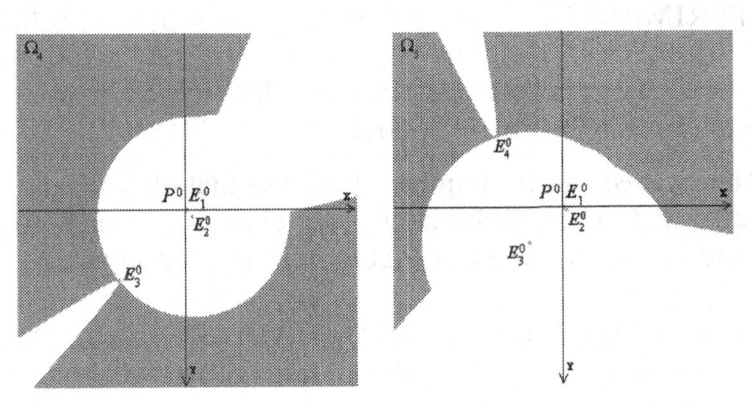

Figure 3.

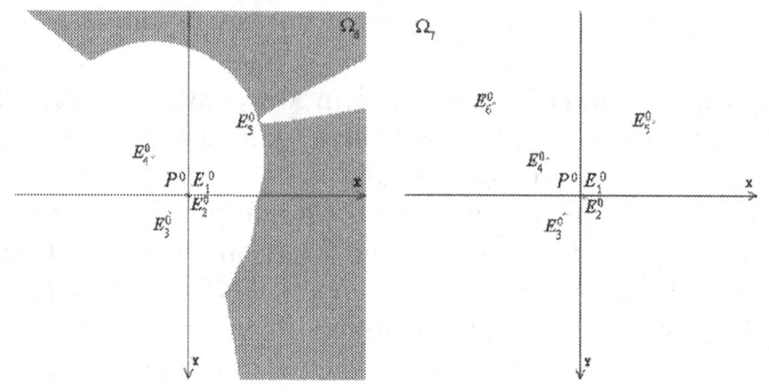

Figure 4.

evader E_7. This means that there is no initial position of evader E_7, at which he would be willing to obey the pursuer.

So, we can conclude that in the game $\Gamma(z_p^0, z_1^0, \ldots, z_7^0)$ the punishment strategy P is not completely realizable as it is not profitable for E_7 to do what the pursuer says. But then in the subgame with six evaders $\Gamma(z_p^0, z_1^0, \ldots, z_6^0)$ the punishment strategy is completely realizable as inequality (3) holds for all E_i $(i = 2, \ldots, 6)$.

We have considered the case when the difference between the players' velocities is large. We also found out that as this difference increases, the realizability areas become wider. Hence, the pursuer can control a larger number of evaders. The following example illustrates this fact.

EXPERIMENT 2. Let $\alpha = 1$, $\beta_1 = \beta_2 = \beta_3 = \beta_4 = 0.999$, $P^0 = (0, 0)$, $E_1^0 = (1, 1)$, $E_2^0 = (-6, -4)$, $E_3^0 = (30, -11)$, $E_4^0 = (x, y)$. Let us construct area Ω_2 like it was done in the previous example (Figure 5(left)).

Here we see that the number of evaders that obey the pursuer is equal to 3. The fourth one will not obey the pursuer. Areas, Ω_3 and Ω_4 are displayed in Figure 5 (right) and Figure 6.

EXPERIMENT 3. Let $m = 2$, $P^0 = (0, 0)$, $E_1^0 = (a, 0)$. Denote by $R = \min_{i=1,\ldots,m} \alpha/\beta_i$ the minimal ratio between the pursuer's speed and that of the evader.

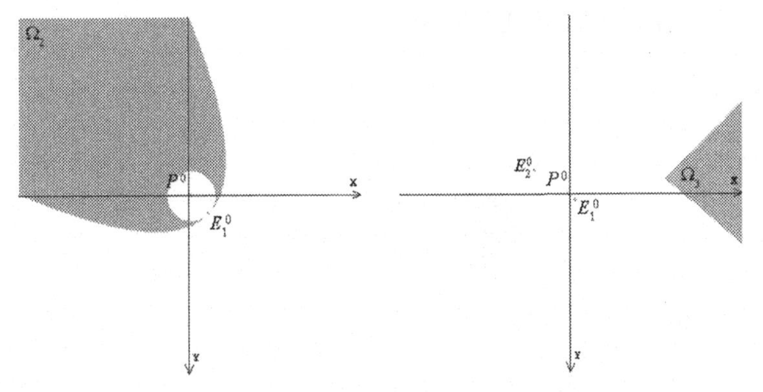

Figure 5.

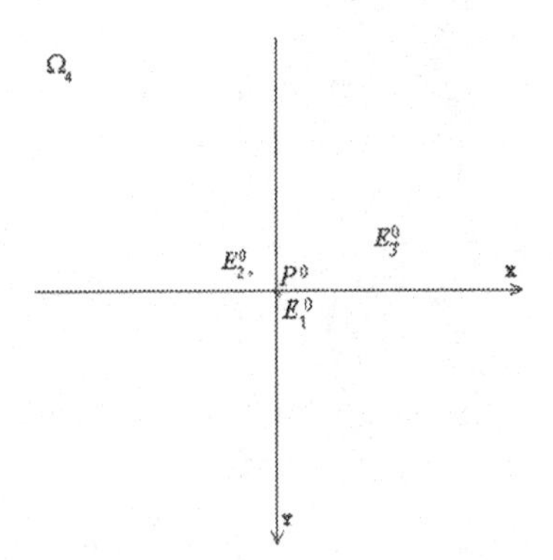

Figure 6.

Consider three cases:

(1) $\alpha = 100,\ \beta_1 = \beta_2 = 1$;

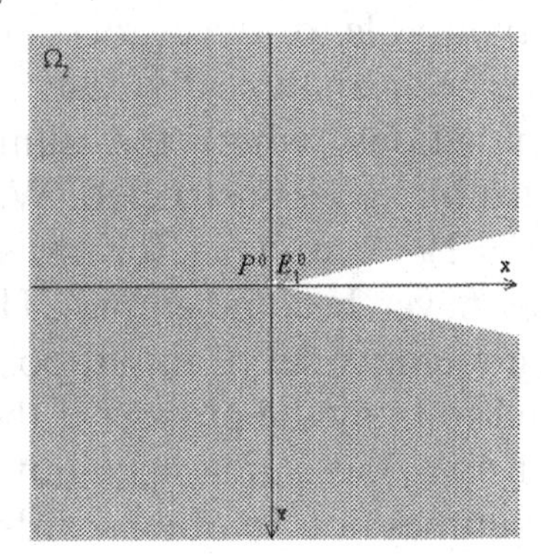

(2) $\alpha = 1$, $\beta_1 = \beta_2 = 0.99999$;

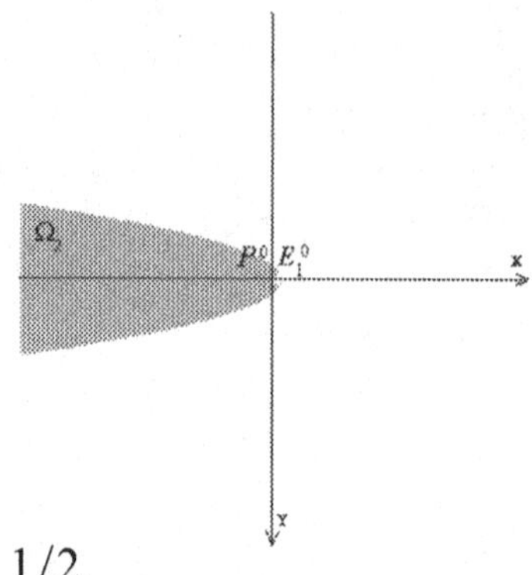

(3) $\alpha = 1$, $\beta_1 = \beta_2 = 1/2$.

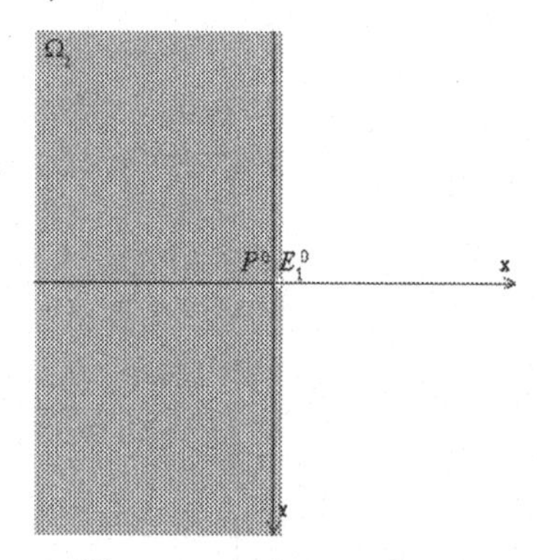

This experiment is to illustrate how the considered area Ω_2 changes with varying $R : R_{(1)} = 100$, $R_{(2)} = 0.00001$, $R_{(3)} = 1/2$.

6. CONCLUSION

In this paper, we have constructed the Nash equilibrium, which is extremely unfavourable for the evaders, and the conditions of its existence have been analytically described. The sets of the evaders' initial positions, where the punishment strategy of pursuer P is realizable, are constructed. We found out that for two evaders ($m = 2$) as well as for three ones ($m = 3$) the realizability areas never become empty. This means that on a plane there is a nonempty set of initial positions for which the evaders are forced to obey the pursuer if they want to be better off. However, for $m > 3$, it can happen not to be so. Sometimes it is true, but sometimes not. In other words, when $m > 3$ (more than three evaders) the number of obedient evaders depends on their initial positions and speeds. Experimentally, we found out that in most of the cases (for most of the checked profiles of players' initial positions) the maximal number of obedient evaders is equal to nine. In fact, keeping a large number of people in submission is practically impossible. As an example, the management of a company has a hierarchical structure based on the fact that each principal is able to control just several agents; otherwise, the industrial process would not be effective.

REFERENCES

Isaacs, R. (1965), *Differential Games: A Mathematical Theory with Applications to Warfare and Pursuit. Control and Optimization.* New York: Wiley.

Owen, G. (1995), *Game Theory*, 3rd edn. San Diego: Academic Press.

Petrosjan, L. and Tomskii, G. (1983), *Geometry of Simple Pursuit.* Moscow: Science.

Tarashnina, S. (1998), *Nash Equilibria in a Differential Pursuit Game with one Pursuer and m Evaders. Game Theory and Applications, Vol. 3.* New York: Nova Science, 115–123.

Address for correspondence: Yaroslavna Pankratova, Svetlana Tarashnina, Faculty of Applied Mathematics and Control Processes, Saint-Petersburg State University, Bibliotechnaya pl. 2, Petrodvorets, Saint-Petersburg, 198504, Russia. E-mails: yasyap@fromru.com, st@isdgrus.ru

ENRICO DENTI and NANDO PRATI

RELEVANCE OF WINNING COALITIONS IN INDIRECT CONTROL OF CORPORATIONS

ABSTRACT. In this paper we study coalitions of indirect stockholders of a company showing that they can have different controlling power, and therefore different relevance in the control problem. We then introduce a suitable classification, and three algorithms to find all the coalitions of all relevances.

JEL Classification: C71, D21.
Mathematics Subject Classification (2000) 91A12.

1. INTRODUCTION

In this paper we study coalitions of direct stockholders or indirect stockholders (briefly SH's) of a given company ***F***, or the similar (but simpler) case of a parliament whose parties are structured in different wings, or other analogous situations. On the contrary of what has been done so far, we take into account the fact that a SH can abstain or can oppose to other SH's. First we discuss the behaviour of a direct or indirect SH showing that not all the winning coalitions have the same controlling power, i.e. the same relevance. Then we extend the algorithm presented in Denti and Prati (2001) so as to calculate all the winning coalitions of all relevances.

2. BASIC DEFINITIONS

Given a firm ***F***, called the *basic firm*, and all its direct or indirect SH's, let ***IS***(***F***) be the set of the (direct or indirect) SH's of ***F*** plus the firm ***F*** itself. The indirect control of ***F*** can be represented (or visualized) by the *indirect shareholding* graph of ***F***,

Theory and Decision **56:** 183–192, 2004.

$\boldsymbol{G}(\boldsymbol{F})$, that is, the directed weighted graph whose set of nodes is $\boldsymbol{IS}(\boldsymbol{F})$ and such that an arc from node X to node Y exists if and only if X owns some stocks of Y, and the arc weight is the real number r representing the percentage of the stocks of firm Y owned by firm X: i.e., if $r = 0.75$, then X owns the 75% of Y. Also denote by $\boldsymbol{T}(\boldsymbol{F})$ the *indirect shareholding table* of $\boldsymbol{F}$, that is, the $n \times n$ table where the value at place X,Y represents the percentage of stocks of the firm Y owned by the firm X.

The assumption that $\boldsymbol{IS}(\boldsymbol{F})$ contains all the direct and indirect SH's assures that the ownership information for each firm is complete, and then the sum in each column of $\boldsymbol{T}(\boldsymbol{F})$ is 1 (i.e. 100%). With the following example we see that in indirect control the behaviour of a SH not belonging to a coalition S greatly influences the possibility of S of being winning or not.

EXAMPLE. Consider the basic firm A: suppose that $\boldsymbol{IS}(\mathrm{A}) = \{\mathrm{A}, \mathrm{B}, \mathrm{C}, \mathrm{D}, \mathrm{E}, \mathrm{F}, \mathrm{G}, \mathrm{H}\}$, and $\boldsymbol{T}(\mathrm{A})$ and $\boldsymbol{G}(\mathrm{A})$ are as in Table I and Figure 1 below.

So the direct SH's of A are B, C, D and the indirect SH's E, F, G, H. We assume throughout the paper that control of a company is achieved with simple majority, but everything works also with other majority quota. Let us consider the game "indirect control of A". The SH's of A can be classified as: (a) *pure investors*, visualized as nodes without entering arcs in the graph and empty columns in the Table I; (b) *companies*, all the other nodes in the graph that correspond to non-empty columns in the table. Here, pure investors are F, G, and H. Note that a pure investor can be dummy (i.e. never necessary to control anything, e.g. H), and an investor, or a SH, can abstain or be absent at a SH's meeting, as it often happens for small SH's in a big company (Figure 1).

At a first sight, players of the game may seem to be only the pure investors, but this is not the case: indeed, all the SH's, including companies, should be considered as players. This is necessary because, even though a company is not free to act as a pure investor, its staff can behave freely between two SH's meetings. So all the SH's must be considered as players.

TABLE I
T (A)

	A	B	C	D	E	F	G	H
A								
B	30		30					
C	30				30			
D	40	10						
E		30						
F		30	30		40			
G		30	40	90	30			
H				10				

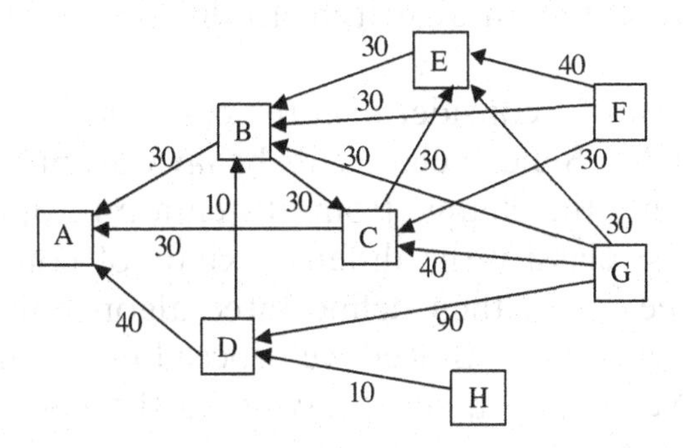

Figure 1. ***G***(A).

Moreover, the freedom that the staff of a company has for some time, can be very useful for the staff itself.

For instance: (1) consider the coalition {C, G}: if all the other SH's abstain for a while (i.e. do nothing in order to prevent the coalition's action) when {C,G} starts moving, *the coalition wins*, since it controls first E, then B, and finally A. After B has been gained to the coalition's control, the situation is blocked: even if someone else tries to oppose, nothing can change. So, the coalition {C,G} could be very useful for the staff of C, which may avoid the risk of being fired by the action of someone else. *Indeed*: {B,F} could well control C if they move first, and, by doing so, also control E, thus closing the game cutting (the staff of) G away. The same may happen if some other coalition moves first.

(2) Take {B,C}: it can control A only if most of the other SH's abstain: otherwise, it can easily be destroyed – for instance, by {F,G}, that can gain control of B and C and then of A.

The indirect control problem has already been studied in literature (see Gambarelli and Owen 1994; Gambarelli, 1996; Denti and Prati, 2001) under the constraints that (i) a SH can only either "try to control the basic firm" or "abstain", and (ii) the behaviour of SH's is fixed in time – i.e., if they begin to try to control the basic firm (abstain), then they will continue to try to control the firm (respectively abstain) at any subsequent time, if they are not compelled by the majority of their SH's to help the coalition. Under these assumptions, in Denti and Prati (2001) we presented an algorithm to determine winning coalitions.

Here we want to consider also the case when a SH can "try to oppose" and a SH can change its behaviour, but we will limit to consider only the change from abstention to opposition. To do so, we first classify the different kinds of controlling relevance in three types, then define three algorithms to find the coalitions of each type. Indeed we proceed in a non-completely formalized way for simplicity leaving to the reader the (easy) task to complete the formalization.

The complete characterisation of the indirect control, taking into account time and all possible changes in the behaviour of SH's, is much more complex and will be discussed in a future paper (for games with abstention and opposition see Prati, 2002).

3. A CLASSIFICATION OF THE CONTROLLING RELEVANCE

As shown in the above example, whether a coalitions wins or not strictly depends on the behaviour of the other SH's – i.e., each coalition has its own relevance in control. Given a firm $\boldsymbol{F}$ and the set $\boldsymbol{IS}(\boldsymbol{F})$, let us consider a coalition S (i.e., $S \subseteq \boldsymbol{IS}(\boldsymbol{F})$). Then we introduce the following definition (note: A−B denotes usual set difference):

DEFINITION

(a) C(S) is the set of firms that *can* be controlled by S.
(b) NC(S) = ***IS***(***F***) – C(S) is the set of firms or investors that *cannot* be controlled by S.
(c) PO(S) = ***IS***(***F***) – *S* is the set of *potential opponents* to S.
(d) S is a *Potentially Winning Coalition* (PWC), if and only if S can get the control of ***F*** in some case.
(e) S is a *Potentially Stably Winning Coalition* (PSWC), if and only if NC(S) is not a PWC.
(f) S is a *Stably Winning Coalition* (SWC), if and only if PO(S) is not a PSWC.

It is easy to see that definitions d, e, f, are equivalent to the following facts:

OBSERVATION. (a) S is a PWC if and only if S can control ***F*** only if all the SH's in PO(S) abstain for a while (indeed this is the best possible situation for S), but can lose if some SH's in PO(S) start opposing.

(b) S is a PSWC if and only if: (1) S can get the control of ***F*** (together with the control of some other firms) only if some of the SH's in PO(S) abstain for a while, and, once the control of ***F*** has been achieved, S maintains it even if some SH in PO(S) decides to oppose. Equivalently: (2) S gets and maintains the control of ***F*** in the best possible situation for S, i.e., when all the other SH's abstain when it begins to move, and they continue to abstain up to the moment when S gets the indirect control of all the firms it can; after that, S continues to keep the control of ***F*** even if the SH's in NC(S) start opposing.

(c) S is a SWC if and only if: (1) S can control ***F*** whichever the behaviour of the SH's in PO(S) is, even in the worst possible situation, i.e. when everyone outside S opposes from the beginning. Or equivalently: (2) NC(PO(S)) is a PWC.

In the classical case all the winning coalitions are SWC: there are no potentially winning or potentially stably-winning coalitions. The following obvious property can be derived too:

PROPOSITION. If a coalition is a SWC, then it is also a PSWC and an PWC; if it is a PSWC, then it is also a PWC. Formally: denoting by SWC, $PSWC$, PWC the sets of SWC, PSWC and PWC coalitions, respectively, then $SWC \subseteq PSWC \subseteq PWC$.
So we can give the following classification:

DEFINITION
(a) $LC = \boldsymbol{IS}(\boldsymbol{F}) - PWC$ is the set of losing coalitions.
(b) $CWC = PWC - PSWC$ is the set of conditionally winning coalitions.
(c) $CSWC = PSWC - SWC$ is the set of conditionally stably winning coalitions.
(d) SWC is the set of stably winning coalitions.

4. THE ALGORITHMS TO COMPUTE WINNING COALITIONS

Hereafter we assume that a basic firm $\boldsymbol{F}$, the set $\boldsymbol{IS}(\boldsymbol{F})$, its table $\boldsymbol{T}(\boldsymbol{F})$, and its graph $\boldsymbol{G}(\boldsymbol{F})$ are given. We report here the algorithm of Denti and Prati (2001) to find all the winning coalitions in the case all the SH's never oppose and their behaviour does not change in time. The algorithm is similar to a methodology already considered in Gambarelli and Owen (1994).

4.1. *The Algorithm for PWC*

Let $\boldsymbol{IS}(\boldsymbol{F}) = \{F_1, \dots, F_n\}$: then, a coalition S of SH's can be formally represented by the n-dimensional vector $\mathrm{Z(S)} = \langle z_1, \dots z_n \rangle$ such that: z_j is 1 if $F_j \in \mathrm{S}$, and 0 otherwise. Then the algorithm works as follows:

ALGORITHM P
Step P,0: Let us take a coalition S along with the vector $Z_0 = Z(\mathrm{S})$ that represents it.
Step P,1: (1) Let $X_1 = Z(\mathrm{S}) \cdot \boldsymbol{T}(\boldsymbol{F}) = \langle x_1^1, \dots, x_n^1 \rangle$ (product of the vector Z(S) by the matrix $\boldsymbol{T}(\boldsymbol{F})$);
(2) Define $Y_1 = \langle y_1^1, \dots, y_n^1 \rangle$ such that $y_j^1 = 1$ if $x_j^1 > 50\%$, 0 otherwise;
(3) Let $Z_1 = \langle z_1^1, \dots, z_n^1 \rangle$ such that $z_j^1 = \max(z_j, y_j^1)$.

These vectors represent: X_1 the number of stocks the coalition S can "move" in order to control some SH directly; Y_1 the SH's that are controlled by S directly; Z_1 the coalition of all the SH's that are now trying to control $\boldsymbol{F}$, i.e. S *plus* the SH's that S has been able to control directly and are now helping S to control $\boldsymbol{F}$. All these SH's are represented by a 1 in these vectors. The algorithm proceeds by induction:

Step P,i+1: Let Z_i represent the SH's that are now trying to control $\boldsymbol{F}$. Then:

(1) Let $X_{i+1} = Z_i \cdot \boldsymbol{T}(\boldsymbol{F}) = \langle x_1^{i+1}, \ldots, x_n^{i+1} \rangle$;

(2) Let $Y_{i+1} = \langle y_1^{i+1}, \ldots, y_n^{i+1} \rangle$ such that $y_j^{i+1} = 1$ if $x_j^{i+1} > 50\%$, 0 otherwise;

(3) Let $Z_{i+1} = \langle z_1^{i+1}, \ldots, z_n^{i+1} \rangle$ such that $z_j^{i+1} = \max(z_j^i, y_j^{i+1})$.

These vectors represent: X_{i+1} the number of stocks that the coalition S can "move" with the help of the other SH's in Z_i; Y_{i+1} the SH's that are controlled directly by the SH's in Z_i, and, in the graph $\boldsymbol{G}(\boldsymbol{F})$, this corresponds to the SH's controlled indirectly by S that can be reached by a path of length i starting from the nodes in S; Z_{i+1} all the SH's that are now trying to control $\boldsymbol{F}$.

It can be easily seen that (as noted in Denti and Prati (2001)), the sequence of vectors Z_i is non-decreasing by construction (i.e., for each index i, $Z_i \leq Z_{i+1}$, which means that, for each j, $z_j^i \leq z_j^{i+1}$), and it becomes also stable (i.e., there is an index n such that, for each $m > 0$, $Z_{n+m} = Z_n$). So, if the vector Z_n has a 1 in the place representing the firm $\boldsymbol{F}$, then $\boldsymbol{F}$ is controlled by S; otherwise it is not controllable by S.

4.2. *The Algorithms for PSWC and SWC*

The above algorithm operates under the hypothesis that all the SH's never oppose and their behaviour does not change in time: this is why it actually calculates all the PWCs, i.e., the greatest set among PWC, PSWC, SWC. Therefore, defining an algorithm to compute PSWC amounts at identifying which coalitions in PWC actually belong to PSWC (subset of PWC) – and analogously for SWC with respect to PSWC The following proposition comes to help:

PROPOSITION. Apply P to a coalition S, and find the vector Z_n where the sequence of vector stabilises. Then Z_n defines also the sets C(S) and NC(S): the SH's in C(S) are represented by a 1 in Z_n, while the SH's in NC(S) are represented by a 0 in Z_n.

This property and the observation in section 3 make it possible to determine *PSWC* and *SWC* by repeatedly applying the above algorithm in the following suitable way. Given a coalition S and the corresponding vector Z(S):

ALGORITHM PS

Step PS,1: apply algorithm P to coalition S. Find the vector Z_n where induction stabilises: according to the above proposition, we have automatically the set NC(S).

Step PS,2: apply algorithm P to the set NC(S), thus determining the stabilised vector Z'_m: if there is a 1 in the place representing ***F*** then NC(S) gets the control of ***F***, and so is a PWC, which means that S itself is not a PSWC; otherwise S is a PSWC.

ALGORITHM S

Step S,1: apply algorithm P to PO(S) until the vector Z_n stabilises: this leads to determine the sets C(PO(S)) and NC(PO(S)) (note: NC(PO(S)) $\subseteq$ S).

Step S,2: apply again algorithm P to the set NC(PO(S)), thus determining the stabilised vector Z'_m: if there is a 1 in the place representing ***F*** then NC(PO(S)) is a PWC, which means that S itself is not a SWC; otherwise S is a SWC.

By these algorithms we can also check whether a coalition S is in *LC*, or *CWC*, or *CSWC*, or *SWC*.

4.3. *The Example (Revised)*

In the example introduced in Section 2, take the coalition $S = \{B, F, G\}$ and the vector representing it $Z_0 = Z(S) = \langle 0, 1, 0, 0, 0, 1, 1, 0 \rangle$. Applying algorithm P to S we obtain: $Z_1 = \langle 0, 1, 1, 1, 1, 1, 1, 0 \rangle$, $Z_2 = \langle 1, 1, 1, 1, 1, 1, 1, 0 \rangle$, where P stabilises. So S is a PWC, and NC(S) = {H}. Applying PS (i.e., applying P to NC(S)) we obtain that NC(S) is not a PWC and so S is at least a PSWC. Consider now its potential opponents PO(S) = {C, D, E, H}: the vector representing PO(S) is $Z(PO(S)) = Z'_0 = \langle 0,0,1,1,1,0,0,1 \rangle$. Now: (1) apply algorithm P to PO(S) until the sequence of vector stabilises: it turns out that $Z_1 = \langle 1,0,1,1,1,0,0,1 \rangle$, where P stabilizes, so NC(PO(S)) = {B, F, G}; (2) apply again algorithm P to the set NC(PO(S)) found above. We find: $Z_0 = \langle 0,1,0,0,0,1,1,0 \rangle$, then $Z_1 = \langle 0,0,1,1,1,1,1,0 \rangle$, and finally $Z_2 = Z_3 = \langle 1,1,1,1,1,1,1,0 \rangle$. Since there is a 1 in the place representing the basic firm A, the coalition S = {B, F, G} is a SWC.

4.4. *Extensive Application of the Algorithm to the Example*

The example is particularly difficult due to the presence of a loop where no firm in the loop is dummy in the control of the subsequent firm in the loop itself. Applying the preceding algorithms to this example we found:
SWC = {{B, C, E, F}, {B, C, E, G}, {B, F, G}, {C, F, G}, {E, F, G} and their supersets},
PSWC = {{B,F},{C,F},{E,F},{B,G},{C,G},{E,G},{F, G} and their supersets}.
PWC = {{B, C}, {B, D}, {B, F}, {C, F}, {E, F}, {B, G},{C, G}, {E, G}, {F, G}, {C, D}, and their supersets}.

REFERENCES

Denti, E. and Prati, N. (2001), An algorithm for winning coalitions in indirect control of corporations, *Decisions in Economics and Finance* 24, 153–158.

Gambarelli, G. (1996), Takeover algorithms. Modelling techniques for financial markets and bank management in M. Bertocchi, E. Cavalli and S. Komlosi (eds), *Proceedings of the 16th and 17th Euro Working Group of Financial Modelling Meeting*, Heidelberg: Physica Verlag, 212–222.

Gambarelli, G. and Owen, G. (1994), Indirect control of corporations, *International Journal of Game Theory* 23, 287–302.

Owen, G. (1995), *Game Theory*, 3rd edn. San Diego, CA: Academic Press.

Prati, N. (2002), Giochi Cooperativi senza Impegno Totale: Una Introduzione. Working Paper Dipartimento Finanza dell' Impresa e dei Mercati Finanziari, Universita' di Udine 2-2002 (in Italian).

Addresses for correspondence: Nando Prati, Dipartimento Finanza dell' Impresa e dei Mercati Finanziari, Universita' di Udine, Via Tomadini 30/a, 33100 Udine, Italia. E-mail: nando.prati@dfimf.uniud.it

Enrico Denti, Dipartimento di Elettronica, Informatica e Sistemistica, Universita' di Bologna, Viale Risorgimento 2, 40136 Bologna, Italia. E-mail: edenti@deis.unibo.it

GIANFRANCO GAMBARELLI and SERENA PESCE

TAKEOVER PRICES AND PORTFOLIO THEORY

ABSTRACT. Takeover operations can have significant repercussions on Portfolio selection as the concentration of shares necessary for the former is in contrast with the diversification implied by the latter. The models which have so far dealt with takeovers have been based on Cooperative Game Theory (more specifically on power indices) integrated with the classical models of Portfolio Theory. There was still a need to perfect a model to forecast the price curve, for use as a benchmark to establish a takeover bid. This paper tackles the problem by means of a non-cooperative approach. The proposed model also makes it possible to consider takeover and portfolio theories from a single viewpoint.

KEY WORDS: Portfolio, takeover, share-holder

1. INTRODUCTION

1.1. *Takeover Models*

Power indices can represent a reasonable expectation of the allotment of coalitional power among the various members of a shareholders' meeting, depending on their shares, see Gambarelli (1993, 1999, 2002) for it. Using these functions as a starting point, formulae have been created which express changes in "coalition power" for each single shareholder, following trading of shares with another shareholder or an ocean of small shareholders (see Gambarelli, 1983, Gambarelli and Szegö, 1982). In this case the power is a step function of the number of acquired shares. The discontinuity points of this function constitute the critical stocks. Several algorithms have been implemented to enable automatic calculation of power indices (see Gambarelli, 1990; Mann and Shapley, 1962) especially with regard to the relative variations ensuing from share trading (see Arcaini and Gambarelli, 1986; Gambarelli, 1996). In this context, quite interesting results regard the evaluation of the risk to the purchaser during the buying and selling of shares

Theory and Decision **56:** 193–203, 2004.

(see Gambarelli, 1993) and the definition of power in indirect control of firms. Indirect control occurs, for example, when an investor owns a quota of shares of a firm which in turn owns shares of another firm and so on. The problem was solved in Gambarelli and Owen, (1994) where a procedure was developed which could transform the set of the different interlinked majority games into a single game. This procedure is based on the concept of 'Multilinear Extensions of Games' introduced by Owen (1972).

1.2. *The Connection Between Takeover and Portfolio Theories*

It is a well-known fact that the classical models of Portfolio Selection advise the saver to diversify his share portfolio in such a way as to efficiently reduce risk (see Szegö, 1980). This, however, is in conflict with the relevant amount of a single stock that needs to be acquired to carry out hostile take-over bids (TOB). The connection between takeover and portfolio theories was initially approached by Amihud and Barnea (1974) and by Batteau in (1980). There were, however, problems in defining the variations in the power quotas held by each shareholder during the buying and selling of shares. The results given in the previous section made it possible to solve this problem. Initially, a control propensity index is defined, which is related to the risk aversion index. On the basis of the first index, control is optimized using power indices. The second index is used to place the residual capital in the purchase of shares as a simple investment, applying classical Portfolio Theory. Furthermore, in this latter context the variance-covariance matrix (see Szegö, 1980) is modified, eliminating data relating to companies involved in the hostile bid (including any linked company). The hostile bid requires a longer time span than the one necessary to obtain a return on a normal investment. The one proposed in Gambarelli (1982) is therefore a multi-period model, based on the progressive movement of investment capital in the classic portfolio aimed at the purchase of the shares necessary for the hostile bid.

1.3. *A Shortcoming*

When merging two such investment models, the question of perfecting the nature of the initial data remains. More precisely, in Portfolio Theory the share purchase price is a variable known a priori. The operations designed to acquire control, on the other hand, are unknown a priori as they depend on the TOB and on any countermoves made by the current controlling group. This paper intends to propose a forecasting model of price trends based on the functions that express the monetary value attributed to control by those two players (raider and controlling group). The different evaluations lead to a process of actions–counter-actions, which is here interpreted in terms of a non-cooperative game. The case is put forward, quite common in real life, where the raider decides to undertake a hostile bid by launching a TOB, and the controlling group reacts by making a counter-offer.

This model is presented in Section 2. In Section 3 its implications will be analyzed through a closer connection between portfolio and takeover theories. The last section deals with the conclusions and gives indications on possible further research on this topic.

2. PRICE FORECASTING

2.1. *The Variables*

With reference to the firm involved and to the relative critical stock, the following notation will be adopted:

p_0 is the official share price.
p^r is the share price offered by the raider in the TOB.
p^c is the share price offered by the current controlling group as a counter-offer in response to the raider's bid.
π^r is the raider's profit in relation to critical stock.
π^c is the current controlling group's profit in relation to critical stock.
m^r is the monetary value of the raider's control.
m^c is the monetary value of the current majority group's control.
q^r is the number of shares needed for the raider to acquire the critical stock.
q^c is the number of shares needed for the current controllers to maintain control.

C^{r} is the total cost to acquire q^{r}.
C^{c} is the total cost to maintain q^{c}.
K^{r} are fixed costs including share purchase at price p_0 by the raider before TOB.
K^{c} are fixed costs for current controlling group.

2.2. *The Model*

In a perfect market the rise in prices, depending on the number of shares bought, coincides with the rise in control power (i.e., with the power index suitably weighted). In an imperfect market, however, these trends are different. It is in this difference that the investor's real advantage lies. A model which forecasts the share price trend as a function of demand is therefore needed.

A model to forecast share prices, based on the different evaluation of control monetary values, is proposed below. Firstly, this is done on the part of the raider and secondly on the part of the current controlling group. In a real situation it often happens that the control of a firm is sought by more than one subject. Consequently, once the raider has demonstrated his intention to acquire control by making a public bid, the current controlling group (or simply an opponent who tries not to lose his share of power) may make a counter-offer.

In the model proposed here, a hypothetical suggestion is made which fixes the optimum price for the raider in contrast with the opponent's choice as the solution of a non-cooperative dynamic game. An adapted version of Stackelberg 2-pole model (see Von Stackelberg, 1952) is used. It is evident that the opponent, when setting the price of his bid, is aware of the raider's choice of strategy. In establishing his bidding price, the raider has to take into consideration his opponent's reaction.

The amount established for the TOB must be attractive enough to make the other shareholders relinquish their shares and, at the same time, sufficient to allow the raider to make an adequate profit. It is therefore necessary to define, by means of estimations of statistical data, the curve of the demand for shares where quantity is a function of the price. More precisely, the number of shares that the raider manages to acquire is directly proportionate to the price of launch established. The same is true for the current controlling group. According to the Stackelberg model, which is adapted here, the curve is shown as a linear combination of the estimated prices of the two opponents. In formulae:

$$q^{r} = a + p^{r} + bp^{c} \quad \text{for the raider}$$
$$q^{c} = a + p^{c} + bp^{r} \quad \text{for the current control group}$$

where a and b are real constants (we simplify the model considering a and b the same in both demand functions; of course formulas for different parameters can be easily obtained). As the model is developed, profit is determined as the difference between the function of the monetary value of control and the overall costs to be paid for the acquisition of the required power position. In symbols:

$$\pi^{r} = m^{r} - C^{r}, \quad \text{where} \quad C^{r} = K^{r} + p^{r}q^{r}$$
$$\pi^{c} = m^{c} - C^{c}, \quad \text{where} \quad C^{c} = K^{c} + p^{c}q^{c}$$

It is therefore a game with perfect information where the two rational decision-makers act in sequence. The raider acts first, forcing the current controlling group to bid second. The solution is achieved as follows.

First, the optimum reaction price p_{m}^{c} is calculated for the second investor (in this case the current controlling group) at the price fixed by the raider p^{r}. This reaction price is calculated by maximizing the profit π^{c} of the current controlling group in relation to the value of the price p^{c}:

$$\max_{p^{c}} \pi^{c} = \max_{p^{c}}(m^{c} - K^{c} - p^{c}(a + p^{c} + bp^{r}))$$

Differentiating the equation above, the following is obtained:

$$\partial\pi^{c}/\partial p^{c} = -a - 2p^{c} - bp^{r}$$

The concavity of the function π^{c} implies that there is only one maximum point p_{m}^{c}, where the derivative equals zero:

$$p_{m}^{c} = -(a + bp^{r})/2 \qquad (1)$$

The raider, in fixing the opening bid p^r, takes into consideration the reaction price of his opponent p^c_m calculated above and maximizes his profit in relation to the opening bid p^r:

$$\max_{p^r} \pi^r = \max_{p^r}(m^r - K^r - p^r(a + p^r + bp^c_m))$$

Replacing p^c_m with the value calculated above, we obtain:

$$= \max_{p^r}(m^r - K^r + ap^r(b/2 - 1) + (p^r)^2(b^2/2 - 1))$$

For $|b| \geq \sqrt{2}$ each point p^r is a maximum point (case $b = \pm\sqrt{2}$ and $a = 0$) or no p^r is a maximum point (as the function to be maximized is an upper unlimited function).

For $|b| < \sqrt{2}$ the function has only one maximum point p^r_m, where

$$\partial\pi^r/\partial p^r = a(b/2 - 1) + p^r(b^2 - 2)$$

$$p^r_m = a(2 - b)/2(b^2 - 2)$$

Replacing the optimum value of p^r in equation (1), the price proposed by the current controlling group is:

$$p^c_m = a(b^2 + 2b - 4)/4(2 - b^2)$$

In the considered interval $|b| < \sqrt{2}$ and for a positive a, we have $p^r_m > p^c_m$.

3. THE MERGER

The forecast for the trend of the price curve makes it possible to develop a model which combines two opposing investment procedures (concentration versus diversification) into a single viewpoint. More precisely, this model calculates the rate of return obtained from the takeover operation and its risk and links the investment to acquire control of a company to 'normal' investments, in order to create the optimal portfolio. The objective is to choose the optimal portfolio which maximizes the profit function of the investor in the long term.

3.1. *Determining Share Capital for the Takeover Operation*

The investor can calculate the overall cost he has to bear for the purchase of the shares required for the takeover operations. In fact, we assume that he owns a forecast model for the price curve trend. The share capital to be allotted for this operation is equal to the value of the overall costs (which, in the case of the raider, is shown as C^r). The allotment of capital according to the two opposing investment procedures, however, no longer occurs as in the previous models on the basis of a propensity parameter of control, but in relation to a criterion of overall optimization. For a creation of the perfect portfolio it is useful to compare the benefits expected from acquiring control of the firm, with the profit ensuing from the advantages of investment diversification.

"Opportunity cost" is intended as the difference between the portfolio performance which also includes the stock of the company being taken over, and the portfolio performance without this stock. It is therefore necessary to check that the profit which can be made by obtaining a majority share is higher than the opportunity cost. Consequently, the investor should make his bid for the company which maximizes the difference between the profit (made both from the majority share in the company and from the portfolio consisting of the remaining available stock) and the opportunity cost.

3.2. *The Computation of the Performance Rate Ensuing from the Takeover Operations*

In the merger model proposed here, the investment to acquire control of a company is compared to a normal investment to create the optimal portfolio. The overall expected performance is in fact achieved from a portfolio which includes:

- the investment in the company to be taken over, and
- the investment in the remaining shares which are negatively correlated.

It is therefore possible to calculate, again in reference to the Shapley–Shubik power index, the performance rate obtained for each power share owned.

The following notation will be used:

υ is the performance rate obtained from the takeover operations

σ_c is the risk involved in the control performance rate

The value of υ is obtained from the relation between the estimate of the value of the profit being made (i.e. the difference between the monetary value of control m^r and the overall costs C^r) and the value of the invested capital C^r. σ_c is a function of the performance rate υ. In symbols

$$v = (m^r - C^r)/C^r$$

$$\sigma_c = f(v)$$

3.3. *The Capitalization of Performance Rates in Normal Investments*

The investment aimed at the takeover operations is therefore included in the creation of the optimum portfolio, considered as a single long-term unit. To equate the performance expected from control with that from other normal investments, it is necessary to standardize the time span accordingly. Since the achievement of the benefits deriving from control requires a greater length of time than the return on a normal investment, we propose to refer all performance rates to the long term, equal to the time needed to benefit from the financial return from the takeover operations.

If t is the time required to benefit from the financial returns on takeover and σ^U is the performance rate of a "normal" (classic) investment, we have to consider the performance rate σ_t^U of the classic investment capitalized for the period t. For instance, in the case of simple (linear) capitalization, we have $\sigma_t^U = t\sigma^U$.

3.4. *Optimization of the Investor's Utility Function*

The risk rate of a portfolio is given by correlation rates of the investments and by the covariances (the covariance is a "weighted" average of the product of the difference between the rates of return and the expected value). According to classic models of Portfolio Selection, it is necessary to use the variance/covariances matrix to solve the optimization problem. At this point it is therefore necessary to exclude from the variance/covariance matrix (see Szegö, 1980) the stock which is closely linked to that of the company to be taken over and to modify the matrix periodically. Changing the values of the matrix means setting up another portfolio. This new portfolio is different from the previous one, either by replacing some stock, or by varying the capital invested in the different companies. Consequently the new portfolio presents a different performance and risk rate which is always able to maximize the function of utility. Use is therefore made of a classic model of portfolio selection which enables us to maximize the raider's utility function. Among the conditions which must be complied with, however, there is a further restriction which limits the capital share for the takeover operations to the value previously established. This makes it possible to achieve the performance rate expected from control of risk σ_c. Basically, the investor includes in his portfolio those investment procedures which will allow him to maximize his benefits. By diversifying he will attempt to absorb the risk assumed through the control of one or more companies, a risk which is usually higher than that implied by other types of investment.

4. CONCLUSIONS

Power index theory allows the investor to define his power of control on the basis of the number of shares owned and to plan a potential takeover bid. Dynamic games, on the other hand, allow us to make a forecast regarding the price curve trend before making a TOB, based on the different evaluations that the various shareholders of a company attribute to the monetary value of control. It is possible to further link the portfolio theory with the takeover operations in order to maximize the earning capacity of invested capital.

Further developments of this research could come from the multi-agent model proposed by Parkes and Huberman (2001). A number of applications of previous models with real data have already been carried out; it is now necessary to test the model proposed here using real data. Finally, further developments could extend this model by dealing with the problem of indirect control of firms, as described in Section 1.1.

ACKNOWLEDGEMENTS

This work is sponsored by MIUR 60%, University of Bergamo.

REFERENCES

Amihud, Y. and Barnea, A. (1974), *Portfolio Selection for Managerial Control.* Omega, 75–83.

Arcaini, G. and Gambarelli, G. (1986), *Algorithm for Automatic Computation of the Power Variations in Share Trading.* Calcolo, 13–19.

Batteau, P. (1980), Approches formelles du problème du contrôle des firmes et des sociétés par actions, *Revue de l'Association Française de Finance* 1, 1–26.

Gambarelli, G. (1982), Portfolio selection and firms' control, *Finance* 3(1), 69–83.

Gambarelli, G. (1983), Common behaviour of power indices, *International Journal of Game Theory* 12(4), 237–244.

Gambarelli, G. (1990), A new approach for evaluating the Shapley value, *Optimization* 21(3), 445–452.

Gambarelli, G. (1993), An index of stability for controlling shareholders, in R. Flavell (ed), *Modelling Reality and Personal Modelling: Proc. of the EURO Working Group of Financial Modelling Meeting*, Heidelberg: Physica Verlag, 116–127.

Gambarelli, G. (1994), Power indices for political and financial decision making: A review, *Annals of Operations Research* 51, 165–173.

Gambarelli, G. (1996), Takeover algorithms, in M. Bertocchi, E. Cavalli and S. Kolomsi (eds), *Modelling Techniques for Financial Markets and Bank Management*, Heidelberg: Physica Verlag, 212–222.

Gambarelli, G. (1999), Minimax apportionments, *Group Decision and Negotiation* 8(6), 441–461.

Gambarelli, G. and Owen, G. (1994), Indirect control of corporations, *International Journal of Game Theory* 23(4), 287–302.

Gambarelli, G. and Owen, G. (2002), Power in political and business structures, in M.J. Holler, H. Kliemt, D. Schnidtchen and M.E. Streit (eds), *Power and Fairness (Jahrbuch für Neue politische Ökonomie), Vol. 20*, Tübingen: Mohr Siebeck pp, 57–68.

Gambarelli, G. and Szegö, G.P. (1982), *Discontinuous Solutions in n-person Games. New Quantitative Techniques for Economic Analysis.* New York: Academic Press, 229–244.

Mann, I. and Shapley, L.S. (1962), *Values of Large Games, VI: Evaluating the Electoral College Exactly*. Santa Monica, CA: RM 3158, Rand Corporation.
Owen, G. (1995), *Game Theory*, 3rd edn. San Diego: Academic Press.
Owen, G. (1972), Mutilinear extensions of games, *Management Science* 18, 76–88.
Parkes, D.C. and Huberman, B.A. (2001), Multiagent cooperative search for portfolio selection, *Games and Economic Behaviour* 35, 124–165.
Von Stackelberg, H. (1952), *The theory of the market economy*. (Translated from the German and with the introduction by A.T. Peacock) London.
Szegö, G. (1980), *Portfolio Theory*. New York: Academic Press.

Address for correspondence: Gianfranco Gambarelli, Department of Mathematics, Statistics, Computer Science and Applications, University of Bergamo, Italy. E-mail: gambarex@unibg.it.

Serena Pesce, Banco di Brescia, Italy.
E-mail: serenapesce@bancalombarda.it

VITO FRAGNELLI

A NOTE ON THE OWEN SET OF LINEAR PROGRAMMING GAMES AND NASH EQUILIBRIA

ABSTRACT. In this paper we associate a strategic non-cooperative game to a linear programming game; we analyze the relations between the core of the given game and the Nash equilibria of the strategic game.

KEY WORDS: Linear programming games, Owen set, Nash equilibrium.

1. INTRODUCTION

In this paper we consider the following situation: some agents own a bundle of resources that can be used for a common project; if a subset of agents agrees on a fair division of the income they develop the project using their resources. This is precisely the case of a linear production game. Now we suppose that before that the agents succeed in carrying out the joint project an external monopolistic company offers to purchase the resources. The company interacts separately with each agent, asking him to communicate the price he wants for his whole bundle of resources. After that the company knows the requirements of the agents decides which bundles to buy, trying to maximize the quantity of resources rather then her gain; more precisely the company will not accept to have a loss (in this case she will not buy the resources) but she is indifferent among different amounts of gain (including the null amount). According to this aim the company accepts to buy those bundles that allow producing goods such that their value is at least equal to the sum of the prices paid. Consequently each agent has to decide which price to ask in order to maximize his utility, i.e. the price should be not too high, so that the company buys the bundle, but it has to be as large as possible.

This problem may be analyzed under the light of game theory. We may represent the starting situation as a

Theory and Decision **56:** 205–213, 2004.

cooperative game where the players are the agents and the characteristic function assigns to each coalition the value that its players may obtain using just their own resources. On the other hand in the final situation the agents may be viewed as players of a non-cooperative game in which the strategies of each player are the possible selling prices he can ask for and his payoff is the selling price he fixed, if his bundle is purchased and is zero if the company thinks that his price is too high (also w.r.t. the prices fixed by the other players), so that the bundle is not sold.

Two questions arise: How to associate a non-cooperative game to a given cooperative game with the same set of players? And is it possible to exploit the characteristics of the cooperative game in order to solve the non-cooperative one?

Many authors tackled the question of associating a non-cooperative game to a cooperative game. Von Neumann and Morgenstern considered the problem of transforming a game in characteristic form into a game in strategic form, in a general framework; Borm and Tijs devoted their interest to cooperative games with non transferable utility and recently Gambarelli proposed a general transformation procedure in which each player have to decide how much to ask for entering in a coalition and his payoff depends on the probability that a coalition in which he takes part forms. Another example of how to define a non-cooperative game is given in Feltkamp et al., where the authors consider a minimum cost spanning tree situation, in which each player has to decide how to spend in the construction of an edge towards an adjacent vertex in order to minimize his cost but succeeding in being connected to the source.

Our setting involve all linear programming games (see Samet and Zemel), i.e. those games in which the players control a bundle of resources and the worth of a coalition is the best output that the players can obtain using just their resources.

In the Section 2 we formalize the problem, in Section 3 we state a connection among the solutions of the cooperative and non-cooperative games and Section 4 concludes.

2. THE INTERTWINED GAMES

A cooperative game is a pair $G^C = (N, v)$, where $N = \{1,\ldots,n\}$ is the set of players and $v : 2^N \rightarrow R$ is the characteristic function that represents the worth that each coalition $S \subseteq N$ can obtain independently from the other players.

In a cooperative game $G^C = (N, v)$ the players have to agree on a division of the worth of the game, i.e. the value of the grand coalition $v(N)$. In order to complete this task a relevant role is played by the core, i.e. the set $core(v) = \{x \in R^n$ s.t. $\sum_{i \in N} x_i = v(N), \sum_{i \in S} x_i \geq v(S) \forall S \subseteq N\}$. In fact an allocation $x \notin core(v)$ does not satisfy the rights of some coalition. If the core is non-empty the game is said to be balanced.

In a linear programming game we have that:

$$v(S) = \max\{c^T y \text{ s.t. } Ay \leq b^S\} \quad \forall S \subseteq N$$

where $c = \{c_1,\ldots,c_m\}$ is the vector of the prices of the goods $y = \{y_1,\ldots,y_m\}$, $A = (a_{lh})_{l=1,\ldots,m;\ h=1,\ldots,k}$ is the matrix of the coefficients of the constraints and the vector $b^S = \Sigma_{i \in S}\, b^i$ represents the resources available for coalition S, where $b^i = \{b^i_1, \ldots, b^i_k\}$ is the vector of the resources owned by each player $i \in N$.

According to a well-known result by Owen the linear programming games are balanced and a core allocation x can be obtained starting from an optimal solution z^* of the dual program $\min\ \{(b^N)^T z \text{ s.t. } A^T z = c;\ z \geq 0\}$ as $x_i = (b^i)^T z^*$, $i \in N$. van Gellekom et al. called *Owen set* the subset of core allocations that are associated, in the previous sense, to dual optimal solutions.

A non-cooperative game is a triple $G^{NC} = (N, (\Sigma_i)_{i \in N}, (\pi_i)_{i \in N})$, where $N = \{1,\ldots,n\}$ is the set of players, Σ_i is the non-empty set of strategies of player $i \in N$ and $\pi_i : \prod_{j \in N} \Sigma_j \rightarrow R$ is the payoff function of player $i \in N$ that defines the payoff of player i for each strategy profile $\sigma = (\sigma_1,\ldots,\sigma_n)$, $\sigma_j \in \Sigma_j$, i.e. when each player $j \in N$ chooses the strategy $\sigma_j \in \Sigma_j$.

A solution of a non-cooperative game is a strategy profile; the most important solution is the Nash equilibrium, i.e. a strategy profile $\sigma^* = (\sigma^*_1, \ldots, \sigma^*_n)$ s.t. no player may improve his payoff if it is the unique player that deviates from his equilibrium strategy, or formally if $\pi_i(\sigma_1^*, \ldots, \sigma_i^*, \ldots, \sigma_n^*) \geq \pi_i(\sigma_1^*, \ldots, \sigma_i, \ldots, \sigma_n^*)$ for each $\sigma_i \in \Sigma_i$, for each $i \in N$.

We already said that the strategies of player $i \in N$ are the prices he may ask for his bundle of resources; we can suppose that he asks for a price at least equal to his worth $v(i)$ and at most equal to the maximum worth of the coalitions which he belongs to, after leaving to the other players their minimum price. This assumption seems reasonable because if player i carries out the project by himself he can anyhow obtain the value $v(i)$ while if he ask for a price larger than the above maximum he surely will not sell his bundle, whatever price the other players ask for their bundles. Formally the price $p(i)$ required by player i has to satisfy

$$v(i) \leq p(i) \leq \max\{v(S) - \sum_{j \in S \setminus \{i\}} v(j), S \ni i\}, \quad \forall i \in N$$

For a profit game we have that

$$\max\{v(S) - \sum_{j \in S \setminus \{i\}} v(j), S \ni i\} = \left\{v(N) - \sum\nolimits_{j \in M(i)} v(j)\right\}$$

as for each $S \subseteq N$ we have

$$v(N) - \sum_{j \in N \setminus \{i\}} v(j) \geq v(S) - \sum_{j \in S \setminus \{i\}} v(j) \Leftrightarrow v(N) \geq v(S) + \sum_{j \in N \setminus S} v(j)$$

that holds by balancedness, so:

$$v(i) \leq p(i) \leq v(N) - \sum_{j \in N \setminus \{i\}} v(j), \quad \forall\, i \in N$$

EXAMPLE 1. Production situation.

Consider the production situation associated to the following technology matrix:

	Units of resource R_1	Units of resource R_2	Value
Item A	3	1	4
Item B	2	1	3

Let $N = \{I, II\}$; the following table describes, for each coalition $S \subseteq N$, the bundle of resources, the items that can be produced at best and the value in the linear production game:

S	R_1	R_2	A	B	$v(S)$
{I}	2	2	0	1	3
{II}	5	1	1	0	4
{I, II}	7	3	1	2	10

The selling prices have to satisfy the following constraints:

$$3 \leq p(I) \leq 6$$
$$4 \leq p(II) \leq 7$$

The non-cooperative game is shown in the following table (only integer values of prices appear):

I / II	4	5	6	7
3	(3, 4)	(3, 5)	(3, 6)	(3, 7)
4	(4, 4)	(4, 5)	(4, 6)	(0, 0)
5	(5, 4)	(5, 5)	(0, 0)	(0, 0)
6	(6, 4)	(0, 0)	(0, 0)	(0, 0)

3. THE CORE AND THE NASH EQUILIBRIA

In this Section we want to study the relation among the core of the cooperative game and the Nash equilibria of the non-cooperative game.

It is easy to check that the core allocations are Nash equilibria; in fact given a core allocation if a player unilaterally deviates from it, he cannot improve his payoff because if he reduces his price, then his payoff trivially decreases, while if he increases his price then his bundle is not sold and his payoff goes down to zero. More precisely if a set of prices $p(1),\ldots, p(n)$ corresponds to a core allocation it satisfies $\Sigma_{j \in S}\, p(j) \geq v(S)$ for each $S \subseteq N$; so if a player $i \in N$ increases his price $p(i)$ then we have $\Sigma_{j \in S}\, p(j) > v(S)$ for each $S \ni i$ and, consequently, the he cannot anymore sell his bundle because for each set of players the sum of the prices is strictly larger than the value of the resources.

We are interested mainly in the Owen set because from a practical point of view it is very easy to find one of such core allocations, simply referring to the dual solution of the linear program associated to the problem whose right hand side is the set of resources of the grand coalition. This process ends in a Nash equilibrium that results to be efficient, i.e. the sum of the payoffs of the players is maximal.

Referring to Example 1 we can observe that $core(v) = \{(\alpha, 10 - \alpha) \in \mathrm{R}^2 \mid 4 \leq \alpha \leq 7\}$ coincides with the set of Nash equilibria. In particular the equilibrium profile (4, 6) corresponds to the dual solution (1, 1) of the production problem related to the grand coalition.

In general the core and the set of Nash equilibria do not coincide, as the following example shows.

EXAMPLE 2. Flow situation.

Consider the flow situation represented by the following network, where $N = \{\mathrm{I, II, III}\}$ (the notation of each arc is *owner/capacity*):

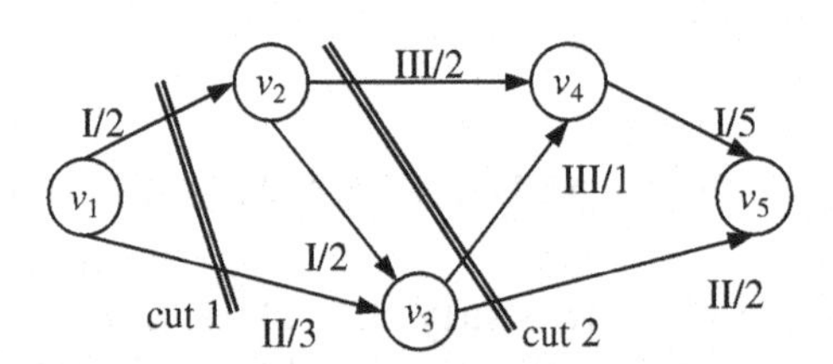

In the associated flow game the players of each coalition $S \subseteq N$ can use only their arcs for maximizing the flow from v_1 to v_5; the resulting game is:

S	{I}	{II}	{III}	{I, II}	{I, III}	{II, III}	N
$v(S)$	0	2	0	2	2	2	5

As in Example 1 we may associate a non-cooperative game; if the players are asked for selling their arcs, the selling prices that players may ask for have to satisfy:

$$0 \leq p(\mathrm{I}) \leq 3$$

$$2 \leq p(\mathrm{II}) \leq 5$$

$$0 \leq p(\mathrm{III}) \leq 3$$

In the following we represent the non-cooperative game where only integer selling prices appear and the Nash equilibria are in bold:

	I / II	2	3	4	5
III = 0					
	0	(0, 2, 0)	(0, 3, 0)	(0, 4, 0)	(0, 5, 0)
	1	(1, 2, 0)	(1, 3, 0)	(1, 4, 0)	(1, 0, 0)
	2	(2, 2, 0)	**(2, 3, 0)**	(2, 0, 0)	(2, 0, 0)
	3	**(3, 2, 0)**	(0, 0, 0)	(0, 0, 0)	(0, 0, 0)
III = 1					
	0	(0, 2, 1)	(0, 3, 1)	(0, 4, 1)	(0, 0, 1)
	1	(1, 2, 1)	**(1, 3, 1)**	(1, 0, 1)	(1, 0, 1)
	2	**(2, 2, 1)**	(0, 0, 0)	(0, 0, 0)	(0, 0, 0)
	3	(0, 2, 0)	(0, 0, 0)	(0, 0, 0)	(0, 0, 0)
III = 2					
	0	(0, 2, 2)	**(0, 3, 2)**	(0, 0, 2)	(0, 0, 2)
	1	**(1, 2, 2)**	(0, 0, 0)	(0, 0, 0)	(0, 0, 0)
	2	(0, 2, 0)	(0, 0, 0)	(0, 0, 0)	(0, 0, 0)
	3	(0, 2, 0)	(0, 0, 0)	(0, 0, 0)	(0, 0, 0)
III = 3					
	0	**(0, 2, 3)**	(0, 0, 0)	(0, 0, 0)	(0, 0, 0)
	1	(0, 2, 0)	(0, 0, 0)	(0, 0, 0)	(0, 0, 0)
	2	(0, 2, 0)	(0, 0, 0)	(0, 0, 0)	(0, 0, 0)
	3	**(0, 2, 0)**	(0, 0, 0)	(0, 0, 0)	(0, 0, 0)

In particular the equilibrium profiles (2, 3, 0) and (0, 2, 3) correspond to the optimal cuts, $\{(v_1, v_2), (v_1, v_3)\}$ and $\{(v_2, v_4), (v_3, v_4), (v_3, v_5)\}$, that are core allocations, according to the result of Kalai–Zemel .

We want to remark that there exists the Nash equilibrium (0, 2, 0) that is not efficient and, consequently, does not correspond to a core allocation.

4. CONCLUDING REMARKS

In this paper we propose how to associate a non-cooperative game to a linear programming game, in a suitable framework. Moreover we point out that a simple way to find a Nash equilibrium for the non-cooperative game is to refer to a dual solution of the linear problem, that leads to a core allocation, in particular to an allocation in the Owen set, as the core is a (strict) subset of the set of Nash equilibria.

REFERENCES

Borm, P. and Tijs, T. (1992), Strategic Claim Games corresponding to an NTU Game, *Games and Economic Behavior* 4, 58–71.

Feltkamp, V., Tijs, T. and Muto, S. (1999), Bird's tree allocations revisited. in García-Jurado, I., Patrone, F., and Tijs, S. (eds.), *Game Practice: Contribution for Applied Game Theory*, Kluwer Academic Publisher: Dordretch, The Netherlands, 75–89.

Gambarelli, G. (2000), Transforming games from characteristic into normal form. Working Paper.

Kalai, E. and Zemel, E. (1982), Totally balanced games and games of flow. *Mathematics of Operations Research* 7, 476–478.

Owen, G. (1975), On the core of linear production games, *Mathematical Programming* 9, 358–370.

Samet, D. and Zemel, E. (1984), On the core and dual set of linear programming games. *Mathematics of Operations Research* 9, 309–316.

Van Gellekom, J.R.G., Potters, J.A.M., Reijnierse, H., Engel M.C. and Tijs, S.H. (2000), Characterization of the Owen set of linear production processes, *Games and Economic Behavior*, 32, 139–156.

Von Neumann, J. and Morgenstern, O. (1944), *Theory of games and economic behavior*, Princeton University Press: Princeton, NJ.

Address for correspondence: Vito Fragnelli, Dipartimento di Scienze e Tecnologie Avanzate, Università del Piemonte Orientale, Piazza Giorgio Ambrosoli, 5, 15100 Alessandria, Italy (Tel.: +39-131-283822; Fax: +39-131-254410; E-mail: vito.fragnelli@mfn.unipmn.it)

N. LLORCA, E. MOLINA, M. PULIDO and J. SÁNCHEZ-SORIANO

ON THE OWEN SET OF TRANSPORTATION SITUATIONS

ABSTRACT. This paper presents an axiomatic characterization of the Owen set of transportation games. In the characterization we use six properties including consistency (CONS2) and splitting and merging (SM) which are firstly proposed and defined for this setup in the present paper.

KEY WORDS: Transportation situations, cooperative games, Owen set

1. INTRODUCTION

In transportation situations, how to obtain the optimal solution for a single decision-maker is well known but a new problem arises when the agents involved cooperate and they have to distribute the obtained profit or saving among each other. In the cooperative game theory literature, we can find many general procedures (solution concepts) for distributing the total amount obtained by a set of agents among each other. In order to select the most suitable procedure of distribution or allocation, it is usual to characterize the solution concepts through different sets of properties. Nowadays, many operations research problems are studied from a game theoretical point of view introducing the very realistic multiple agents component in the analysis of such problems. For a survey on this interesting topic see Borm et al. (2002). In particular, transportation situations are studied from a game theoretical point of view in Samet et al. (1984), Sánchez-Soriano (1998) and Sánchez-Soriano et al. (2001) among others.

The purpose of this paper is to investigate certain aspects of transportation games and give an axiomatic characterization of its Owen set (a solution concept related to the dual optimal solutions of the transportation problem). To achieve this objective, we look at two operations research problems closely

Theory and Decision **56:** 215–228, 2004.

related to transportation situations: assignment situations and linear production situations. Assignment situations and linear production situations were studied from a game theoretical point of view in Shapley and Shubik (1972) and Owen (1975) respectively. A very interesting property in assignment games is that the core and the Owen set of the game coincide. This fact does not occur in linear production situations and transportation situations. Sasaki (1995) provided two characterizations of the core of assignment games and hence of its Owen set, and van Gellekom et al. (2000) provided a characterization of the Owen set of linear production games. Transportation situations occupy an intermediate position in between assignment and linear production situations because assignment situations are a special case of transportation situations and these can be rewritten as linear production situations. This particular position and relationship make us look into the properties used to characterize the Owen set of assignment games and linear production games to find the set of properties that characterize the Owen set of transportation games.

This paper consists of three sections. In the next section we present the most relevant definitions and results for the transportation situations and related games. We study the properties of the Owen set in Section 3. We consider two types of properties which are related to those introduced by Sasaki (1995) to characterize the core of assignment games, and by van Gellekom et al. (2000) to characterize the Owen set of linear production games. In the final section, we provide a characterization of the Owen set in transportation situations.

2. TRANSPORTATION SITUATIONS AND GAMES

A transportation situation is determined by a tuple $T = (P, Q, \mathbf{B}, \mathbf{p}, \mathbf{q})$, where P and Q are, respectively, the sets of supply points and demand points. The transport of one unit of the goods from supply point i to demand point j generates a non-negative profit of b_{ij}. The matrix $\mathbf{B} = [b_{ij}]_{i \in P, j \in Q}$ contains all the profits per unit of the goods. The supply at point $i \in P$ equals p_i units of the goods and the demand $j \in Q$ is q_j units where both p_i and q_j are non-negative integers. The vectors $\mathbf{p} = (p_i)_{i \in P}$ and $\mathbf{q} = (q_j)_{j \in Q}$ contain, respectively, the supplies and demands of the goods. The set of transportation situations is denoted by G^T.

A transportation plan for $T \in G^T$ is a matrix $\boldsymbol{\mu} = [\mu_{ij}]_{i \in P, j \in Q}$ where $\mu_{ij} \geq 0$ is the number of units of the goods that will be transported from supply point i to demand point j. A supply point $i \in P$ can supply at most p_i units of the goods, and a demand point $j \in Q$ wants to receive at most q_j. The maximal profit that can be obtained in this situation is

$$\max\left\{ \sum_{(i,j) \in P \times Q} b_{ij}\mu_{ij} | \boldsymbol{\mu} \text{ is a transportation plan} \right\}.$$

A transportation plan $\boldsymbol{\mu}$ is also called a feasible solution for the above transportation program T. We denote by $O_p(T)$ the set of optimal solutions for this program. Our interest is how to distribute the total profit among the agents when they cooperate, i.e., to propose a vector $(\mathbf{x}; \mathbf{y}) \in \Re^P \times \Re^Q$ such that

$$\sum_{i \in P} x_i + \sum_{j \in Q} y_j = \sum_{(i,j) \in P \times Q} b_{ij}\mu^*_{ij},$$

where $\boldsymbol{\mu}^* \in O_p(T)$.

A well-known approach to distribute the total profit is to define a game for each transportation situation. Given $T \in G^T$, the corresponding *transportation game* (N, w) is a cooperative transferable utility (TU) game with player set $N = P \cup Q$. Let $S \subset N, S \neq \emptyset$, be a coalition of players and define $P_S = P \cap S$ and $Q_S = Q \cap S$. If $S = P_S$ or $S = Q_S$, there are either only suppliers or demanders present, then no transport can take place and the worth $w(S)$ of coalition S equals zero. Otherwise, the worth $w(S)$ depends upon the transportation plans for this coalition. A transportation plan $\boldsymbol{\mu}(S)$ for coalition S is a transportation plan for the transportation problem $T_S = (P_S, Q_S, [b_{ij}]_{i \in P_S, j \in Q_S}, (p_i)_{i \in P_S}, (q_j)_{j \in Q_S})$. By convention $w(\emptyset) = 0$.

Now, when thinking about how to share the profit among the suppliers and the demanders, one can consider sharing it according to some game theoretic solution concept for TU-games. For instance, one way is to do so according to an element in the *core* of the transportation game (N, w), i.e.,

$$C(w) = \left\{ (\mathbf{x};\mathbf{y}) \in \Re^P \times \Re^Q \;\middle|\; \begin{array}{l} \sum_{i\in P} x_i + \sum_{j\in Q} y_j = w(N) \quad \text{and} \\ \sum_{i\in P_S} x_i + \sum_{j\in Q_S} y_j \geq w(S), \quad \text{for all } S\subset N, S\neq\varnothing \end{array} \right\}.$$

If a core-element is proposed as a distribution of the total profit $w(N)$, then each coalition S will get at least as much as it can obtain on its own, therefore no coalition has an incentive to split off. The transportation games have non-empty core and are superadditive but not convex. Thus, for an arbitrary transportation game we can always select a core element, but with great difficulty. Owen (1975) introduced the class of linear production games and presented a method to find a non-empty subset of the core of these games. Since a transportation game can be seen as a special case of linear production games, we can use this method to derive core elements. This set is the so-called *Owen set* of the transportation situation, which is defined by

$$\text{Owen}(T) = \left\{ (\mathbf{x};\mathbf{y}) \in \Re^P \times \Re^Q \;\middle|\; \begin{array}{l} x_i = p_i u_i, \forall i \in P, \;\; y_j = q_j v_j, \forall j \in Q, \\ \text{with } (\mathbf{u};\mathbf{v}) \in O_d(T) \end{array} \right\},$$

where $O_d(T)$ is the set of optimal solutions of the dual program of the (relaxed) transportation program for the grand coalition, i.e.,

$$\min\left\{ \sum_{i\in P} p_i u_i + \sum_{j\in Q} q_j v_j \,\middle|\, u_i + v_j \geq b_{ij}, \;\; u_i, v_j \geq 0, \text{ for all } i\in P, j\in Q \right\}.$$

Note that each element $u_i, i \in P, v_j, j \in Q$, of a vector $(\mathbf{u};\mathbf{v}) \in O_d(T)$ is the mean profit that an agent will obtain per unit from the supply or demand. Therefore, an element $(p_i u_i, q_j v_j)_{i\in P, j\in Q}$ of the Owen set is a vector of profits that players receive from the supply or demand.

Assignment situations are a special case of transportation situations when all supplies $p_i, i \in P$, and demands $q_j, j \in Q$, equal 1. Hence, an assignment situation can be represented as $A = (P, Q, \mathbf{B})$. Assignment games were introduced by Shapley and Shubik (1972). They proved that the core of an assignment game is the non-empty set of optimal solutions of the dual program of the (relaxed) assignment program for the grand coalition, i.e., $C(w) = O_d(A) = \text{Owen}(A)$. In the sequel, we will denote the set of assignment situations by G^A and the core of the assignment game $C(w)$ by $C(A)$.

In general, the Owen set does not coincide with the core of a transportation game. See Sánchez-Soriano et al. (2001) for a detailed analysis of the relationship between the core and the Owen set of a transportation situation.

A *solution concept* Ψ on G^T is a map, which assigns to every $T = (P, Q, \mathbf{B}, \mathbf{p}, \mathbf{q}) \in G^T$ a non-empty set of available distributions $\emptyset \neq \Psi(T) \subset \Re^P \times \Re^Q$.

3. PROPERTIES

This section is devoted to the properties of the Owen set. We begin with those properties which are common to general games, or, at least to two-sided market games. We end up with the definition of two new properties which are related to those introduced by Sasaki (1995) to characterize the core of assignment games, and by van Gellekom et al. (2000) to characterize the Owen set of linear production games.

DEFINITION 1. A solution Ψ on G^T satisfies:

(EFF) Efficiency, if $x(P) + y(Q) = w^T(P \cup Q)$, $\forall(\mathbf{x}; \mathbf{y}) \in \Psi(T), \forall T \in G^T$.

(IR) Individual Rationality, if $(\mathbf{x}; \mathbf{y}) \geq 0, \forall(\mathbf{x}; \mathbf{y}) \in \Psi(T)$, $\forall T \in G^T$.

(CR) Couple Rationality, if $x_i + y_j \geq w^T(\{i,j\}) = b_{ij} \min\{p_i, q_j\}, \forall i \in P, \forall j \in Q, \forall(\mathbf{x}; \mathbf{y}) \in \Psi(T), \forall T \in G^T$.

(CONT) Continuity, if for every $(P, Q, \mathbf{B}, \mathbf{p}, \mathbf{q}) \in G^T$ and every sequence $\{T^k = (P, Q, \mathbf{B}^k, \mathbf{p}, \mathbf{q})\}_{k \in \mathbb{N}}$ of transportation problems and elements $\{(\mathbf{x}^\mathbf{k}; \mathbf{y}^\mathbf{k}) \in \Psi(T^k)\}_{k \in \mathbb{N}}$, such that: (a) $\mathbf{B}^\mathbf{k} \rightarrow \mathbf{B}$ (i.e. $b_{ij}^k \rightarrow b_{ij}$, for all $i \in P, j \in Q$); (b) $(\mathbf{x}^\mathbf{k}; \mathbf{y}^\mathbf{k}) \rightarrow (\mathbf{x}; \mathbf{y})$, it follows that $(\mathbf{x}; \mathbf{y}) \in \Psi(T)$.

PROPOSITION 1. *The Owen Set satisfies* (EFF), (IR), (CR) *and* (CONT) *on* G^T.

Proof. Since every allocation in the Owen set is a core element, it is obvious that the Owen set satisfies (EFF), (IR) and (CR). To prove the continuity of the Owen set, let $\{T^k = (P, Q, \mathbf{B^k}, \mathbf{p}, \mathbf{q})\}_{k\in\mathbb{N}}$, $\{(\mathbf{x}^k; \mathbf{y}^k) \in \text{Owen}(T^k)\}_{k\in\mathbb{N}}$, and let $T = (P, Q, \mathbf{B}, \mathbf{p}, \mathbf{q})$ and $(\mathbf{x}; \mathbf{y})$, such that (i) $\mathbf{B^k} \xrightarrow[k\to\infty]{} \mathbf{B}$ and (ii) $(\mathbf{x^k}; \mathbf{y^k}) \xrightarrow[k\to\infty]{} (\mathbf{x}; \mathbf{y})$.

$(\mathbf{x^k}; \mathbf{y^k}) \in \text{Owen}(T^k)$, $\forall k \in \mathbf{N}$ then there exists $(\mathbf{u^k}; \mathbf{v^k}) \in O_d(T^k)$ such that

$$x_i^k = p_i u_i^k, \ \forall i \in P \quad \text{and} \quad y_j^k = q_j v_j^k, \ \forall j \in Q, \forall k \in \mathbb{N}. \tag{1}$$

Then, it follows from condition (ii) that

$$u_i^k = \frac{x_i^k}{p_i} \xrightarrow[k\to\infty]{} \frac{x_i}{p_i}, \quad \forall i \in P \quad \text{and} \quad v_j^k = \frac{y_j^k}{q_j} \xrightarrow[k\to\infty]{} \frac{y_j}{q_j}, \quad \forall j \in Q. \tag{2}$$

Let $(\mathbf{u}; \mathbf{v})$ given by $u_i = x_i/p_i, \forall i \in P$, and $v_j = y_j/q_j, \forall j \in Q$. We will show that $(\mathbf{u}; \mathbf{v}) \in O_d(T)$.

$(\mathbf{u^k}; \mathbf{v^k}) \in O_d(T^k), \forall k \in \mathbb{N}$, then $(\mathbf{u^k}; \mathbf{v^k}) \geq \mathbf{0}$, and $u_i^k + v_j^k \geq b_{ij}^k, \forall i \in P, \forall j \in Q$. Therefore, taking (i) into account, it follows that $(\mathbf{u}; \mathbf{v}) = \lim_{k\to\infty}(\mathbf{u^k}; \mathbf{v^k}) \in F_d(T)$, where $F_d(T)$ is the feasible set of the dual program of the (relaxed) transportation program for the grand coalition.

Now, in order to prove the optimality of $(\mathbf{u}; \mathbf{v})$, we will find a feasible solution $\boldsymbol{\mu}$ for the transportation program T (i.e. $\boldsymbol{\mu} \in F_p(T)$) with the same value for the objective function.

Since $(\mathbf{u^k}; \mathbf{v^k}) \in O_d(T^k), \forall k \in \mathbb{N}$, then for all $k \in \mathbb{N}$, there exists $\boldsymbol{\mu}^\mathbf{k} \in O_p(T^k) \subset F_p(T^k)$ such that

$$\sum_{i\in P}\sum_{j\in Q} b_{ij}^k \mu_{ij}^k = \sum_{i\in P} p_i u_i^k + \sum_{j\in Q} q_j v_j^k.$$

Note that the feasible set $F_p(T^k) = F_p(T)$, for all $k \in \mathbb{N}$, and $F_p(T)$ is compact. Then, the sequence $\{\boldsymbol{\mu}^k\}_{k\in\mathbb{N}} \subset F_p(T)$ has a convergent subsequence $\{\boldsymbol{\mu}^{k_\lambda}\}_{\lambda\in\mathbb{N}}$. Let $\boldsymbol{\mu}^\mathbf{0} \in F_p(T)$ its limit, then on the one hand

$$\lim_{\lambda\to\infty}\left(\sum_{i\in P} p_i u_i^{k_\lambda} + \sum_{j\in Q} q_j v_j^{k_\lambda}\right) = \lim_{\lambda\to\infty}\sum_{i\in P}\sum_{j\in Q} b_{ij}^{k_\lambda}\mu_{ij}^{k_\lambda} = \sum_{i\in P}\sum_{j\in Q} b_{ij}\mu_{ij}^{0}. \tag{3}$$

On the other hand,

$$\lim_{\lambda\to\infty}\left(\sum_{i\in P} p_i u_i^{k_\lambda} + \sum_{j\in Q} q_j v_j^{k_\lambda}\right) = \sum_{i\in P} p_i u_i + \sum_{j\in Q} q_j v_j. \tag{4}$$

Thus, from (3) and (4) it follows that $(\mathbf{u};\mathbf{v}) \in O_d(T)$. Then, $(\mathbf{x};\mathbf{y}) \in \mathrm{Owen}(T)$. □

The following property was introduced by Sasaki (1995) to characterize the core on assignment problems. Since the core coincides with the Owen set in the assignment problems, we will look at it in order to define a new consistency property for transportation situations. Then, we will use this new consistency to characterize the Owen set.

DEFINITION 2. A solution Φ on G^A satisfies consistency (CONS), if for all $A = (P, Q, \mathbf{B}) \in G^A$, for all $(\mathbf{x};\mathbf{y}) \in \Phi(A)$ and for all $(\overline{P}, \overline{Q}) \subset (P, Q)$, such that there exists a feasible assignment $\boldsymbol{\mu} \in F_p(A)$ with

(i) $x(P) + y(Q) = \sum_{(i,j)\in P\times Q} b_{ij}\mu_{ij}$,
(ii) $\mu_{ij} = 0$, for all pair $(i,j) \in (P\backslash\overline{P}) \times \overline{Q}$ or $(i,j) \in \overline{P} \times (Q\backslash\overline{Q})$,
(iii) $\sum_{i\in\overline{P}} x_i + \sum_{j\in\overline{Q}} y_j = \sum_{(i,j)\in\overline{P}\times\overline{Q}} b_{ij}\mu_{ij}$,

it follows that $(\mathbf{x}|_{\overline{P}}; \mathbf{y}|_{\overline{Q}}) \in \Phi(\overline{P}, \overline{Q}, \mathbf{B}|_{\overline{P}\cup\overline{Q}})$.

The following theorem can be found in Sasaki (1995).

THEOREM 1. *Let Φ a solution on G^A which satisfies* (CONS) *and* (CONT). *If $\Phi(A) \subset C(A)$, for all $A \in G^A$, then $\Phi \equiv C$ on G^A.*

The Owen set satisfies (CONS) on G^A (because it coincides with the core on that class), but not on G^T. Instead of (CONS) the Owen set satisfies on G^T a slightly different version of the consistency property.

DEFINITION 3. A solution Ψ on G^T satisfies (CONS2), if for all $T = (P, Q, \mathbf{B}, \mathbf{p}, \mathbf{q}) \in G^T$, for all $(\mathbf{x}; \mathbf{y}) \in \Psi(T)$, and for all $(\overline{P}, \overline{Q}) \subset (P, Q)$, such that

$$\sum_{i \in \overline{P}} x_i + \sum_{j \in \overline{Q}} y_j = w(\overline{P} \cup \overline{Q})$$

it verifies that $(\mathbf{x}|_{\overline{P}}; \mathbf{y}|_{\overline{Q}}) \in \Psi(\overline{P}, \overline{Q}, \mathbf{B}|_{\overline{P} \cup \overline{Q}}, \mathbf{p}|_{\overline{P}}, \mathbf{q}|_{\overline{Q}})$.

PROPOSITION 2. *The Owen set satisfies* (CONS2) *on* G^T

Proof. Let $T = (P, Q, \mathbf{B}, \mathbf{p}, \mathbf{q}) \in G^T$, and let $(\mathbf{x}; \mathbf{y}) \in \text{Owen}(T)$ and $(\overline{P}, \overline{Q}) \subset (P, Q)$, satisfying the above condition.

Since $(\mathbf{x}; \mathbf{y}) \in \text{Owen}(T)$, there exists $(\mathbf{u}; \mathbf{v}) \in O_d(T)$ such that $x_i = p_i u_i, \forall i \in P$ and $y_j = q_j v_j, \forall j \in Q$.

Clearly, $(\mathbf{u}|_{\overline{P}}; \mathbf{v}|_{\overline{Q}})$ belongs to $F_d(T_{\overline{P} \cup \overline{Q}})$ and

$$\sum_{i \in \overline{P}} p_i u_i + \sum_{j \in \overline{Q}} q_j v_j = \sum_{i \in \overline{P}} x_i + \sum_{j \in \overline{Q}} y_j = w(\overline{P} \cup \overline{Q}),$$

so $(\mathbf{u}|_{\overline{P}}; \mathbf{v}|_{\overline{Q}}) \in O_d(T_{\overline{P} \cup \overline{Q}})$. Therefore, $(\mathbf{x}|_{\overline{P}}; \mathbf{y}|_{\overline{Q}}) \in \text{Owen}(\overline{P}, \overline{Q}, \mathbf{B}|_{\overline{P} \cup \overline{Q}}, \mathbf{p}|_{\overline{P}}, \mathbf{q}|_{\overline{Q}})$. □

The following lemma will be useful in the axiomatic characterization of the Owen set.

LEMMA 1. *Let* Ψ *a solution on* G^T *satisfying* (EFF) *and* (CONS2). *Then* Ψ *satisfies* (CONS) *on* G^A.

Proof. Let $A = (P, Q, \mathbf{B}) \in G^A$, $(\mathbf{x}; \mathbf{y}) \in \Psi(A)$ and $(\overline{P}, \overline{Q}) \subset (P, Q)$ satisfying the conditions in Definition 2. By efficiency and condition (i), the matching $\boldsymbol{\mu}$ is an optimal assignment for A. And hence, by condition (ii), the matching $\boldsymbol{\mu}|_{\overline{P} \times \overline{Q}}$ is an optimal assignment for $A|_{\overline{P} \cup \overline{Q}}$. Therefore, by condition (iii), it is satisfied that

$$\sum_{i \in \overline{P}} x_i + \sum_{j \in \overline{Q}} y_j = w(\overline{P} \cup \overline{Q}).$$

Since Ψ satisfies (CONS2), it follows that $(\mathbf{x}|_{\overline{P}}; \mathbf{y}|_{\overline{Q}}) \in \Psi(\overline{P}, \overline{Q}, \mathbf{B}|_{\overline{P} \cup \overline{Q}})$. □

To each transportation situation $T=(P,Q,\mathbf{B},\mathbf{p},\mathbf{q})\in G^T$ a *representatives assignment situation* $A^T=(P^T,Q^T,\mathbf{B}^T)\in G^A$ can be associated by splitting every supply point $i\in P$ into p_i supplier representatives and every demand point $j\in Q$ into q_j demander representatives. In this way, we will have a set of suppliers $P^T=\{(ir),i\in P,1\leq r\leq p_i\}$ and another set $Q^T=\{(jc),j\in Q,1\leq c\leq q_j\}$ of demanders. The profit generated by matching a supplier (ir) and a demander (jc) is $b^T_{ir,jc}:=b_{ij}$. Clearly, the corresponding operations research problems in both situations are equivalent, as are the optimal solutions.

The following property adapts the property of *shuffle* (van Gellekom et al., 2000) to the context of transportation situations. Note that our property is much less restrictive. We restrict ourselves to integer divisions and unions and we do not ask for permutation invariance. In our context, the following property can be interpreted in terms of a non-manipulability condition.

DEFINITION 4. A solution Ψ on G^T satisfies Splitting and Merging (SM) if for every $T\in G^T$ the following conditions are satisfied:

(S) If $(\mathbf{x};\mathbf{y})\in\Psi(T)$, then there exists $(\mathbf{x}^{A^T};\mathbf{y}^{A^T})\in\Psi(A^T)$ such that

$$x_i=\sum_{r=1}^{p_i}x^{A^T}_{ir},\ \forall i\in P,\quad y_j=\sum_{c=1}^{q_j}y^{A^T}_{jc},\ \forall j\in Q. \tag{5}$$

(M) If $(\mathbf{x}^{A^T};\mathbf{y}^{A^T})\in\Psi(A^T)$, then the distribution $(\mathbf{x};\mathbf{y})\in\Psi(T)$, where

$$x_i=\sum_{r=1}^{p_i}x^{A^T}_{ir},\ \forall i\in P,\quad y_j=\sum_{c=1}^{q_j}y^{A^T}_{jc},\ \forall j\in Q. \tag{6}$$

PROPOSITION 3. *The Owen Set satisfies* (SM).

Proof. We begin by proving the splitting part. We outline the proof as follows. Let $T\in G^T$ and $(\mathbf{x};\mathbf{y})\in\mathrm{Owen}(T)$, then there exists $(\mathbf{u};\mathbf{v})\in O_d(T)$ such that $x_i=p_iu_i$, for all $i\in P$, and $y_j=q_jv_j$, $\forall j\in Q$.

Let $A^T\in G^A$ the assignment situation derived from T. It can easily be checked that $(\mathbf{u^A};\mathbf{v^A})\in O_d(A^T)$, where

$$u_{ir}^{A} = u_i = \frac{x_i}{p_i}, \ \forall i \in P, \ 1 \leq r \leq p_i,$$

$$v_{jc}^{A} = v_j = \frac{y_j}{q_j}, \ \forall j \in Q, \ 1 \leq c \leq q_j.$$

Then, the allocation $(\mathbf{x}^{A^T}; \mathbf{y}^{A^T})$ defined as

$$x_{ir}^{A} = u_{ir}^{A} = \frac{x_i}{p_i}, \ \forall i \in P, \ 1 \leq r \leq p_i,$$

$$y_{jc}^{A} = v_{jc}^{A} = \frac{y_j}{q_j}, \ \forall j \in Q, \ 1 \leq c \leq q_j,$$

belongs to Owen(A^T), with $x_i = \sum_{r=1}^{p_i} x_{ir}^{A}$, for all $i \in P$, and $y_j = \sum_{c=1}^{q_j} y_{jc}^{A}$, for all $j \in Q$.

For the merging condition, let $(\mathbf{x}^{A^T}; \mathbf{y}^{A^T}) \in$ Owen(A^T). Notice that for an assignment situation, $A \in G^A$, Owen$(A) = O_d(A)$. It can easily be checked that the vector $(\bar{\mathbf{u}}; \bar{\mathbf{v}})$ of averages prices, given by

$$\bar{u}_i = \frac{\sum_{r=1}^{p_i} x_{ir}^{A^T}}{p_i}, \ \forall i \in P, \quad \bar{v}_j = \frac{\sum_{c=1}^{q_j} y_{jc}^{A^T}}{q_j}, \ \forall j \in Q,$$

belongs to $O_d(T)$. Therefore, the allocation $(\mathbf{x}; \mathbf{y})$ given by $x_i = p_i \bar{u}_i$, $\forall i \in P$, and $y_j = q_j \bar{v}_j, \forall j \in Q$, belongs to Owen$(T)$ and

$$x_i = \sum_{r=1}^{p_i} x_{ir}^{A^T}, \ \forall i \in P, \quad y_j = \sum_{c=1}^{q_j} y_{jc}^{A^T}, \ \forall j \in Q. \qquad \square$$

4. AXIOMATIC CHARACTERIZATION

Now, we offer an axiomatization involving the properties analyzed in the previous section. We will show that the set of axioms we have considered are logically independent.

THEOREM 2. *There is a unique solution on G^T that satisfies* (EFF), (IR), (CONS2), (CR), (SM), *and* (CONT), *and it is the Owen set.*

Proof. Let Ψ solution on G^T that satisfies (EFF), (IR), (CONS2), (CR), (SM), and (CONT).

First, we will prove that $\Psi \subset$ Owen$(\cdot)$. Let $T \in G^T$, and $(\mathbf{x};\mathbf{y}) \in \Psi(T)$. Since Ψ satisfies (SM), then it follows from the splitting condition that

$$x_i = \sum_{r=1}^{p_i} x_{ir}^{A^T},\ \forall i \in P,\ y_j = \sum_{c=1}^{q_j} y_{jc}^{A^T}, \forall j \in Q,$$

for some $(\mathbf{x}^{A^T};\mathbf{y}^{A^T}) \in \Psi(A^T)$, where A^T is the assignment situation derived from the transportation situation T. By (IR), it follows that

$$\begin{aligned} x_{ir}^{A^T} &\geq v^{A^T}(\{ir\}) = 0, \quad \forall i \in P,\ \forall r = 1,\ldots,p_i, \\ y_{jc}^{A^T} &\geq v^{A^T}(\{jc\}) = 0, \quad \forall j \in Q,\ \forall c = 1,\ldots,q_j. \end{aligned} \tag{7}$$

On the other hand, because Ψ satisfies (CR),

$$\begin{aligned} &x_{ir}^{A^T} + y_{jc}^{A^T} \geq w^{A^T}(\{ir,jc\}) = b_{ir,jc} = b_{ij}, \\ &\quad \forall i \in P,\ \forall j \in Q,\ 1 \leq r \leq p_i,\ 1 \leq c \leq q_j. \end{aligned} \tag{8}$$

Then, it follows from (7) and (8) that $(\mathbf{x}^{A^T};\mathbf{y}^{A^T}) \in F_d(A^T)$. Moreover, since Ψ satisfies (EFF), $(\mathbf{x}^{A^T};\mathbf{y}^{A^T}) \in O_d(A^T) =$ Owen(A^T).

Since Owen$(\cdot)$ satisfies (SM), then by the merging condition it is verified that $(\mathbf{x};\mathbf{y})$, with

$$x_i = \sum_{r=1}^{p_i} x_{ir}^{A^T},\ \forall i \in P,\quad y_j = \sum_{c=1}^{q_j} y_{jc}^{A^T},\ \forall j \in Q,$$

belongs to Owen(T).

Next, we will prove that Owen$(\cdot) \subset \Psi$. We have just proved that every solution satisfying the above properties is a subsolution of the Owen set. Because the Owen set coincides with the core on G^A, then $\Psi(A) \subset C(A)$, for all $A \in G^A$. Moreover, by Lemma 1, Ψ satisfies (CONS) on G^A, and by hypotheses, Ψ also satisfies (CONT). Therefore, by Theorem 1, $\Psi(A) = C(A) =$ Owen(A), for all $A \in G^A$.

Let $(\mathbf{x};\mathbf{y}) \in$ Owen(T), which satisfies (SM). Then, it follows from the merging condition that there exist $(\mathbf{x}^{A^T};\mathbf{y}^{A^T}) \in$ Owen(A^T) verifying (5).

Since $\Psi(A^T) = \text{Owen}(A^T)$, it follows that $(\mathbf{x}^{A^T}; \mathbf{y}^{A^T}) \in \Psi(A^T)$. Thus, taking that Ψ satisfies the merging condition into account, it is verified that $(\mathbf{x}; \mathbf{y}) \in \Psi(T)$. □

PROPOSITION 4. *The axioms* (EFF), (IR), (CR), (CONS2), (SM) *and* (CONT) *are logically independent.*

Proof. To show the independence of these six axioms we will prove that for every group of five axioms there exists a solution which satisfies them all except the sixth one.

($\neg$EFF) Let $\Psi^1(T) = \{(\mathbf{x}; \mathbf{y}) = (p_i u_i, q_j v_j)_{i \in P, j \in Q} | (\mathbf{u}; \mathbf{v}) \in F_d(T)\}$, for all transportation situation $T \in G^T$. Then, the solution concept Ψ^1 trivially satisfies all axioms except efficiency.

Note that every distribution $(\mathbf{x}; \mathbf{y}) = (p_i u_i, q_j v_j)_{i \in P, j \in Q}$ with $u_i + v_j \geq b_{ij}$, for all $(i,j) \in P \times Q$ satisfies couple rationality. In fact,

$$x_i + y_j \geq p_i u_i + q_j v_j \geq \min\{p_i, q_j\} b_{ij} = w(\{i,j\}), \quad \text{for all } i \in P, \ j \in Q.$$

The splitting and merging property is trivially accomplished because the distributions in Ψ^1 only relay on the linear programming program of the underlying transportation problem, which is invariant under the operations of merging and splitting. With respect to the consistency property, the restriction to a coalition $S \subset N = P \cup Q$ of a feasible solution for the dual program of the (relaxed) transportation program for the grand coalition is also feasible for the corresponding dual program of the (relaxed) transportation program for coalition S.

The proof for continuity resembles that of proposition 1 to a high degree.

($\neg$IR)The solution concept on G^T, Ψ^2, defined as

$$\Psi^2(T) = \left\{ (\mathbf{x}; \mathbf{y}) \in \Re^{P \cup Q} \middle| \begin{array}{l} \sum_{i \in P} x_i + \sum_{j \in Q} y_j = w(N), \ \ x_i = p_i u_i, y_j = q_j v_j, \ \forall (i,j) \in P \times Q, \\ \text{with} (\mathbf{u}; \mathbf{v}) \in \Re^{P \cup Q} \text{ s.t. } u_i + v_j \geq b_{ij}, \forall (i,j) \in P \times Q \end{array} \right\},$$

satisfies all axioms except individual rationality. Note that the non-negativity condition of the dual program has been ignored, whereas the efficiency condition and couple rationality have been imposed. With respect to consistency, continuity and splitting and merging, the argument is similar to that Ψ^1 considered above, taking into account that efficiency for coalition S is a necessary condition to ask for consistency.

($\neg$CR) The solution concept on G^T, Ψ^3, defined as

$$\Psi^3(T) = \left\{ (\mathbf{x};\mathbf{y}) \in \Re^{P\cup Q} \;\middle|\; \begin{array}{l} \sum_{i\in P} x_i + \sum_{j\in Q} y_j = w(N), \quad x_i = p_i u_i, \\ y_j = q_j v_j, u_i \geq 0, v_j \geq 0, \forall (i,j) \in P \times Q \end{array} \right\},$$

satisfies all axioms except couple rationality, since the inequalities $u_i + v_j \geq b_{ij}$, $i \in P, j \in Q$ of the dual program have been ignored, whereas the efficiency condition and individual rationality have been imposed. With respect to consistency, continuity and splitting and merging, the argument is similar to that of Ψ^2 considered above.

($\neg$CONS2) The solution concept on G^T, Ψ^4, defined as

$$\Psi^4(T) = \begin{cases} (p_1 b_{11}, 0), & \text{if } |P| = |Q| = 1 \text{ and } p_1 = q_1 \\ \text{Owen}(T), & \text{otherwise} \end{cases}$$

satisfies all axioms except consistency.

Let us consider the transportation problem T with two suppliers and two demanders, supply vector $\mathbf{p} = (1, 2)$, demand vector $\mathbf{q} = (1, 1)$ and transportation matrix $\mathbf{B} = \begin{pmatrix} 10 & 5 \\ 5 & 0 \end{pmatrix}$. The distribution $(\mathbf{x};\mathbf{y}) = (5, 0; 5, 0) \in \Psi^4(T)$ and is efficient for coalition $S = \{1, 3\}$, but $(\mathbf{x}^S;\mathbf{y}^S) = (5; 5) \notin \Psi^4(T_S) = \{(10; 0)\}$.

($\neg$SM) If we remove the splitting and merging property then the core satisfies the remaining properties. We refer to Sánchez-Soriano (1998) for a proof of the continuity.

($\neg$CONT) If we eliminate continuity a solution concept which chooses the relative interior of the Owen set invalidates the result.

ACKNOWLEDGMENTS

This research is supported by Oficina de Ciència i Tecnologia de la Generalitat Valenciana, through project GV-CTIDIA-2002-32 and by Government of Spain, through a joint research grant Universidad Miguel Hernández-Università degli Studi di Genova HI2002-0032.

REFERENCES

Borm, P., Hamers, H. and Hendrickx, R. (2001), Operations research games: A survey, *TOP* 9, 139–216.

Owen, G. (1975), On the core of linear production games, *Mathematical Programming* 9, 358–370.

Samet, D., Tauman, Y. and Zang, I. (1984), An application of the Aumann–Shapley prices for cost allocation in transportation problems, *Mathematics of Operations Research* 9, 25–42.

Sánchez-Soriano, J. (1998), *El problema del transporte. Una aproximación desde la Teoría de Juegos* (in Spanish). Ph.D. Thesis, University of Murcia, Murcia, Spain.

Sánchez-Soriano, J., López, M.A. and García-Jurado, I. (2001), On the core of transportation games, *Mathematical Social Sciences* 41, 215–225.

Sasaki, H. (1995), Consistency and monotonicity in assignment problems, *International Journal of Game Theory* 24, 373–397.

Shapley, L.S. and Shubik, S. (1972), The assignment game I: The core, *International Journal of Game Theory* 1, 111–130.

van Gellekom, J.R.G., Potters, J.A.M., Reijnierse, J.H., Engel M.C. and Tijs, S.H. (2000), Characterization of the Owen set of linear production processes, *Games and Economic Behavior* 32, 139–156.

Address for correspondence: Natividad Llorca, Dpto. Estadística y Matemática Aplicada and C.I.O., Universidad Miguel Hernández de Elche Edificio Torretamarit. Elche, 03202, Spain. E-mail: nllorca@umh.es

Elisenda Molina, Manuel A. Pulido, Joaquín Sánchez-Soriano, Dpto. Estadística y Matemática Aplicada and C.I.O., Universidad Miguel Hernández de Elche, Elche, 03202, Spain E-mail: {e.molina, manpul, joaquin}@ umh.es

E. ALGABA, J.M. BILBAO, J.R. FERNÁNDEZ and A. JIMÉNEZ

THE LOVÁSZ EXTENSION OF MARKET GAMES

ABSTRACT. The multilinear extension of a cooperative game was introduced by Owen in 1972. In this contribution we study the Lovász extension for cooperative games by using the marginal worth vectors and the dividends. First, we prove a formula for the marginal worth vectors with respect to compatible orderings. Next, we consider the direct market generated by a game. This model of utility function, proposed by Shapley and Shubik in 1969, is the concave biconjugate extension of the game. Then we obtain the following characterization: The utility function of a market game is the Lovász extension of the game if and only if the market game is supermodular. Finally, we present some preliminary problems about the relationship between cooperative games and combinatorial optimization.

KEY WORDS: Owen extension, Lovász extension, market games

1. INTRODUCTION

The Owen extension of a cooperative game is introduced by Owen (1972) as a multilinear polynomial in n real variables, where the coefficients are the dividends of the coalitions. The main applications of this extension were obtained with the construction of formulas for the Shapley and Banzhaf values. The Lovász extension of a convex game satisfies an optimization property on the core of the game. We show that this property implies a new formula for the Lovász extension of every cooperative game. Moreover, the coefficients of this extension are also the dividends of the coalitions. Market games and totally balanced games are the same ones and the utility function of the traders in a direct market is continuous and concave. In this paper we prove that the utility function is the Lovász extension of the game if and only if the game is supermodular.

Let us briefly outline the contents of our contribution. In the next section, we provide definitions and preliminaries results on

Theory and Decision **56:** 229–238, 2004.

cooperative game theory. In Section 3 we define market games and some of their classical properties are described. Section 4 is devoted to introduce the Lovász extension of a cooperative game and the link between duality for convex–concave functions and duality for submodular–supermodular functions. The precise statement was proved by Lovász (1983): A set function is submodular if and only if its Lovász extension is convex. Section 4 also includes the representation of Lovász extension using dividends. Section 5 offers the characterization of the utility function of the traders in a direct market. The final section contains an outline of the implications and consequences for game theory of this new link between cooperative game theory and combinatorial optimization.

2. PRELIMINARIES

A cooperative game (N, v) is a function $v : 2^N \rightarrow R$ with $v(\emptyset) = 0$. The players are the elements of the finite set $N = \{1, \ldots, n\}$, and the coalitions are the subsets $S \subseteq N$. To every coalition S is associated its characteristic vector e^S, defined by

$$(e_i^S)_{i \in N} = \begin{cases} 1 & \text{if } i \in S, \\ 0 & \text{if } i \notin S. \end{cases}$$

This is the natural correspondence between 2^N and $\{0,1\}^n$. Through this identification of coalitions with their characteristic vectors, a cooperative game is a function $v : \{0,1\}^n \rightarrow R$ with $v(\mathbf{0}) = 0$.

Unanimity games are considered in order to obtain several extensions of a cooperative game in terms of its dividends. For any $S \subseteq N$, $S \neq \emptyset$,

$$u_S(T) = \begin{cases} 1 & \text{if } T \supseteq S, \\ 0 & \text{otherwise} \end{cases}$$

is called the S-unanimity game. Every game v is a linear combination of unanimity games, that is,

$$v = \sum_{S \subseteq N} a_S(v) u_S,$$

where the coefficients $\{a_S(v) : S \subseteq N, S \neq \emptyset\}$ are the *dividends* of the coalition S in the game v. Any cooperative game (N, v) has a unique expression as a multilinear polynomial in n discrete variables:

$$v(x) = \sum_{S \subseteq N} \left(a_S(v) \prod_{i \in S} x_i \right), \quad x \in \{0, 1\}^n.$$

This polynomial expression using real variables was introduced in game theory by Owen (1972) as the *multilinear extension* $f\colon [0, 1]^n \to R$, of the cooperative game $v : \{0, 1\}^n \to R$. Then, we have $f(e^S) = v(S)$ for all $S \subseteq N$. Owen showed that the Shapley and Banzhaf values are given by

$$\Phi_i(N, v) = \int_0^1 \frac{\partial f}{\partial x_i}(t, \ldots, t)\, \mathrm{d}t,$$

$$\beta_i(N, v) = \frac{\partial f}{\partial x_i}\left(\frac{1}{2}, \ldots, \frac{1}{2}\right), \quad 1 \leq i \leq n.$$

Let us recall some classical solution concepts for cooperative games. The *core* of a game (N, v) is the set

$$\mathrm{Core}(v) = \{x \in R^n : x(N) = v(N), x(S) \geq v(S) \text{ for all } S \subset N\},$$

where for any $S \subseteq N, x(S) = \sum_{i \in S} x_i$ and $x(\emptyset) = 0$.

Let us assume a total ordering of the elements of N, defined by $i_1 < i_2 < \cdots < i_n$. Given the previous ordering C, consider the following chain of coalitions:

$$C_0 \subset C_1 \subset \cdots \subset C_{n-1} \subset C_n,$$

where $C_0 = \emptyset$ and $C_k = \{i_1, i_2, \ldots, i_k\}$, $k = 1, \ldots, n$. The *marginal worth vector* $a^C \in R^n$ with respect to the ordering C in the game (N, v) is given by

$$a^C_{i_k} = v(C_k) - v(C_{k-1}), \quad k = 1, \ldots, n.$$

The *Weber set* of the game (N, v) is the convex hull of the marginal worth vectors, i.e., $\mathrm{Weber}(v) = \mathrm{conv}\{a^C : C$ is an *ordering of* $N\}$. Convex games were introduced by Shapley (1971). A game (N, v) is convex if for every $S, T \in 2^N$,

$$v(S \cup T) + v(S \cap T) \geq v(S) + v(T).$$

It is easy to prove that $a^C(C_k) = v(C_k)$ for $k = 1, \ldots, n$. Weber (1988) showed that any game satisfies $\text{Core}(v) \subseteq \text{Weber}(v)$ and Ichiishi (1981) proved that if $\text{Weber}(v) \subseteq \text{Core}(v)$ then (N, v) is a convex game. Thus, these results imply the following characterization of convex games.

THEOREM 1. *The cooperative game* (N, v) *is convex if and only if* $\text{Core}(v) = \text{Weber}(v)$.

3. MARKET GAMES

We consider a game model of a pure exchange economy. Let $N = \{1, \ldots, n\}$ be the set of traders and we suppose that they participate in a market encompassing trade in m commodities. The space R^m_+ is considered as the commodity space. Every trader $i \in N$ is characterized by means of an initial endowment vector $w^{(i)} \in R^m_+$ and by a utility function $u_i : R^m_+ \to R$ which measures the worth, for him, of any bundle of commodities. The individual utility functions u_i are continuous and concave. The four components vector $\mathbf{M} = (N, m, \{w^{(i)}\}_{i \in N}, \{u_i\}_{i \in N})$ is called a *market*.

We denote by $w(S) = \sum_{i \in S} w^{(i)} \in R^m_+$ the aggregate endowment of the coalition of traders S. Then, $w(S)$ can be reallocated as a collection $\{a^{(i)} : i \in S\}$ of bundles such that each $a^{(i)} \in R^m_+$ and $a(S) = \sum_{i \in S} a^{(i)} = w(S)$. Denote by $A(S)$ the set of these collections. Since the individual utility functions are continuous and $A(S)$ is a compact set, we can define a cooperative game $(N, v_{\mathbf{M}})$ as

$$v_{\mathbf{M}}(S) = \max\left\{ \sum_{i \in S} u_i(a^{(i)}) : \{a^{(i)} : i \in S\} \in A(S) \right\},$$

for all $S \subseteq N$. This model is called a *market game* (see Kannai, 1992), and it corresponds to the original market in a natural way.

A game (N, v) is *totally balanced* if for all $S \subseteq N$ and all e^S-balanced collection, i.e., $\{\gamma_T\}_{T \subseteq N}$ with $\sum_{T \subseteq N} \gamma_T e^T = e^S$ and $\gamma_T \geq 0$, satisfies $\sum_{T \subseteq N} \gamma_T v(T) \leq v(S)$. The following result is due to Shapley and Shubik (1969).

THEOREM 2. *A game is a market game if and only if it is totally balanced.*

4. THE LOVÁSZ EXTENSION

We now consider an extension of a cooperative game (N, v) to a function on R^n. We say that a non-negative function $f : 2^N \to R_+$ is a weighted chain if the family $\Omega = \{S \subseteq N : f(S) > 0\}$ is a chain, i.e., $S \subseteq T$ or $T \subseteq S$ for every pair $S, T \in \Omega$. To every weighted chain f, we associate a non-negative vector

$$x = \sum_{S \subseteq N} f(S) e^S \in R^n_+$$

called the depth vector of f. This is a one-to-one correspondence: for a non-negative vector $x \in R^n_+$ let $0 \leq x^1 < x^2 < \cdots < x^k$ be the components of x with different value and let $S_p = \{i \in N : x_i \geq x^p\}$. Then we define

$$f_x(S) = \begin{cases} x^p - x^{p-1} & \text{if } S = S_p, \\ 0 & \text{otherwise} , \end{cases}$$

where $x^0 = 0$. Obviously f_x is a weighted chain and its depth vector is x.

Let (N, v) be a cooperative game. There is a natural way of extending $v : 2^N \to R$ to all non-negative real vectors.

DEFINITION 1. Let $x \in R^n_+$ be a non-negative vector and f_x its weighted chain, the Lovász extension of v is $\hat{v} : R^n_+ \to R$ defined by $\hat{v}(x) = \sum_{S \subseteq N} f_x(S) v(S)$.

The function $\hat{v}$ is an extension of v because $\hat{v}(e^S) = v(S)$ for all $S \subseteq N$ and it has the following properties:

1. $\hat{v}$ is positively homogeneous, i.e., $\hat{v}(\lambda x) = \lambda \hat{v}(x)$ for all $\lambda \geq 0$.
2. $\widehat{v_1 + v_2} = \widehat{v_1} + \widehat{v_2}$.
3. $\widehat{\lambda v} = \lambda \hat{v}$ for all $\lambda \in R$.

Lovász (1983) proved the following characterization of the supermodular functions.

THEOREM 3. *A function $v : 2^N \to R$ is supermodular if and only if the Lovász extension $\hat{v}$ of v is concave.*

Note that convex games are supermodular functions with $v(\emptyset) = 0$. Moreover, the Lovász extension of a supermodular function satisfies (see Fujishige, 1991) the next optimization property: $\hat{v}(x) = \min\{\langle x, y\rangle : y \in P(v)\}$ for all $x \in R^n_+$, where

$$P(v) = \{y \in R^n : y(S) \geq v(S) \text{ for all } S \subseteq N\}.$$

The Lovász extension is strongly related to the greedy algorithm. An ordering $i_1 < i_2 < \cdots < i_n$ is compatible with the vector $x \in R^n_+$ if $x_{i_1} \geq x_{i_2} \geq \cdots x_{i_n} \geq 0$. We obtain a formula for the marginal worth vectors with respect to x-compatible orderings in convex games.

THEOREM 4. *Let (N, v) be a convex game and let $x \in R^n_+$. The marginal worth vector a^C with respect to an x-compatible ordering C satisfies*

$$\hat{v}(x) = \min\{\langle x, y\rangle : y \in \mathrm{Core}(v)\} = \langle x, a^C\rangle.$$

Proof. If v is a convex game then Theorem 1 implies that $a^C \in Core(v) \subset P(v)$ and hence we obtain the result if we prove $\langle x, y\rangle \geq \langle x, a^C\rangle$ for all $y \in P(v)$. If $y \in P(v)$ then the summation by parts implies that

$$\begin{aligned}
\langle x, y\rangle &= \sum_{k=1}^{n} x_{i_k} y_{i_k} = \sum_{k=1}^{n-1} \left[(x_{i_k} - x_{i_{k+1}}) \sum_{j=1}^{k} y_{i_j} \right] + x_{i_n} \sum_{j=1}^{n} y_{i_j} \\
&\geq \sum_{k=1}^{n-1} (x_{i_k} - x_{i_{k+1}}) v(C_k) + x_{i_n} v(C_n) \\
&= \sum_{k=1}^{n} x_{i_k} [v(C_k) - v(C_{k-1})] \\
&= \sum_{k=1}^{n} x_{i_k} a^C_{i_k} = \langle x, a^C\rangle.
\end{aligned}$$

□

Our next theorem gives the formula for computing the Lovász extension using dividends.

THEOREM 5. *Let* (N, v) *be a game with dividends* $\{a_S(v) : S \subseteq N,\ S \neq \emptyset\}$. *Then the Lovász extension of* v *satisfies* $\hat{v}(x) = \sum_{S \subseteq N} a_S(v) \min_{i \in S} x_i$, *for all* $x \in R^n_+$.

Proof. Properties 2 and 3 of the Lovász extension imply that $\hat{v} = \sum_{S \subseteq N} a_S(v) \widehat{u_S}$. Every unanimity game u_S is a convex game and we can use the optimization property showed in Theorem 4. Thus,

$$\begin{aligned} \widehat{u_S}(x) &= \min\{\langle x, y\rangle : y \in \mathrm{Core}(u_S)\} \\ &= \min\{\langle x, e_i\rangle : i \in S\} = \min_{i \in S} x_i, \end{aligned}$$

where the second equation follows from the characterization of the core for unanimity games (see Einy and Wettstein, 1996):

$$\mathrm{Core}(u_S) = \mathrm{Weber}(u_S) = \mathrm{conv}\{e_i : i \in S\}. \qquad \square$$

REMARK. Driessen and Rafels (1999) have studied several characterizations for the Lovász extension of k-convex games.

5. THE UTILITY FUNCTION

For every totally balanced game (N, v) we consider the following direct market $\mathbf{M}_0 = (N, n, \{e^i\}_{i \in N}, u)$, where u is the same utility function for all traders, given by

$$u(x) = \max\left\{ \sum_{T \subseteq N} \gamma_T v(T) : \{\gamma_T\}_{T \subseteq N} \text{ is an } x\text{-balanced collection} \right\}.$$

This utility funcion is the concave biconjugate extension $v^{\circ\circ}$ of the game v (see Fujishige, 1991). Every convex game is totally balanced. In this case we obtain the following property of the Lovász extension.

THEOREM 6. *Let* (N, v) *be a totally balanced game with common utility function* $u : R^n_+ \to R$. *Then* (N, v) *is convex if and only if its Lovász extension* $\hat{v} = u$.

Proof. The utility function is the solution of the next linear programming problem: $u(x) = \max\{\langle \gamma, v\rangle : A\gamma = x,\ \gamma \geq 0\}$, where $\gamma = (\gamma_T)_{T\subseteq N}$, $v = (v(T))_{T\subseteq N}$ and the matrix $A = (e_i^T)_{i\in N, T\subseteq N}$. The dual problem is

$$u(x) = \min\{\langle x, y\rangle : y^T A^T \geq v\} = \min\{\langle x, y\rangle : y \in P(v)\}.$$

If the game (N, v) is convex then v is supermodular and the optimization property implies $\hat{v} = u$. Conversely, if the utility function satisfies $u = \hat{v}$ then $\hat{v}$ is concave and Theorem 3 implies that (N, v) is a convex game. □

6. OPEN PROBLEMS

The optimization of non-linear functions over the core of a submodular or supermodular function was considered by Fujishige (1991) who extended the Fenchel's duality theory. Also, he introduced the concept of subdifferential of a submodular function. A theory for discrete optimization (non-linear integer programming), called *discrete convex analysis* was developed by Murota (1998). This theory includes functions defined on the integral points in the core of a game.

Some problems about the new relationship between cooperative games and combinatorial optimization are:

1. To apply the Fenchel min–max duality theory and Murota's discrete convex analysis to game theory.
2. To introduce the subgradient vectors and subdifferential sets (see Fujishige, 1984) in cooperative game theory.
3. To study the relationships between the Shapley and Banzhaf values of a game and the subdifferential of its Lovász extension.

Martínez-Legaz (1996) introduced the *indirect function* $\pi : R^n \to R$ of a cooperative game (N, v) by $\pi(x) = \max\{v(S) - x(S) : S \subseteq N\}$, for all $x \in R^n$. Note that the utility function u of the direct market induced by a totally balanced game satisfies

$$u(x) = \min\{\langle x, y\rangle : y \in P(v)\} = \min\{\langle x, y\rangle : y \in \pi^{-1}(0)\}.$$

We propose the following two additional questions for market games:

4. Is there another extension with the utility function property for totally balanced games?
5. To study the relationship between the indirect function of a game and its Lovász extension.

ACKNOWLEDGMENTS

This research has been partially supported by the Spanish Ministery of Science and Technology, under grant SEC2003-00573.

REFERENCES

Driessen, T.S.H. and Rafels, C. (1999), Characterization of k-convex games, *Optimization* 46, 403–431.

Einy, E. and Wettstein, D. (1996), Equivalence between bargaining sets and the core in simple games, *International Journal of Game Theory* 25, 65–71.

Fujishige, S. (1984), Theory of submodular programs: a Fenchel-type min–max theorem and subgradients of submodular functions, *Mathematical Programming* 29, 142–155.

Fujishige, S. (1991), *Submodular Functions and Optimization.* Amsterdam: North-Holland.

Ichiishi, T. (1981), Supermodularity: applications to convex games and to the greedy algorithm for LP, *Journal of Economic Theory* 25, 283–286.

Kannai, Y. (1992), The core and balancedness in Aumann, R.J. and Hart, S. (eds), *Handbook of Game Theory*, Vol. I, Amsterdam: North-Holland, 355–395.

Lovász, L. (1983), Submodular functions and convexity, in Bachem, A. Gröstschel, M. and Korte, B. (eds.), *Mathematical Programming: The State of the Art*, Berlin: Springer-Verlag, 235–257.

Martínez-Legaz, J.E. (1996), Dual representation of cooperative games based on Fenchel-Moreau conjugation, *Optimization* 36, 291–319.

Murota, K. (1998), Discrete convex analysis, *Mathematical Programming* 83, 313–371.

Owen, G. (1972), Multilinear extension of games, *Management Science* 18, 64–79.

Shapley, L.S. (1971), Cores of convex games, *International Journal of Game Theory* 1, 11–26.

Shapley, L.S. and Shubik, M. (1969), On market games, *Journal of Economic Theory* 1, 9–25.

Weber, R.J. (1988), Probabilistic values for games, in Roth, A.E. (ed.), *The Shapley Value*, Cambridge: Cambridge: University Press, 101–119.

Addresses for correspondence: Jesús Mario Bilbao, Department of Applied Mathematics II, Escuela Superior de Ingenieros, Camino de los Descubrimientos, 41092 Sevilla, Spain (E-mail: mbilbao@us.es).

Encarnación Algaba, Julio Rodrigo Fernández, Andrés Jiménez, Department of Applied Mathematics II, Escuela Superior de Ingenieros, Camino de los Descubrimientos, 41092 Sevilla, Spain E-mails: {ealgaba,julio}@ us.es, hispan@matinc.us.es).

THEORY AND DECISION LIBRARY

SERIES C: GAME THEORY, MATHEMATICAL PROGRAMMING AND OPERATIONS RESEARCH

Editor: H. Peters, *Maastricht University, The Netherlands*

1. B.R. Munier and M.F. Shakun (eds.): *Compromise, Negotiation and Group Decision.* 1988 ISBN 90-277-2625-6
2. R. Selten: *Models of Strategic Rationality.* 1988 ISBN 90-277-2663-9
3. T. Driessen: *Cooperative Games, Solutions and Applications.* 1988 ISBN 90-277-2729-5
4. P.P. Wakker: *Additive Representations of Preferences.* A New Foundation of Decision Analysis. 1989 ISBN 0-7923-0050-5
5. A. Rapoport: *Experimental Studies of Interactive Decisions.* 1990 ISBN 0-7923-0685-6
6. K.G. Ramamurthy: *Coherent Structures and Simple Games.* 1990 ISBN 0-7923-0869-7
7. T.E.S. Raghavan, T.S. Ferguson, T. Parthasarathy and O.J. Vrieze (eds.): *Stochastic Games and Related Topics.* In Honor of Professor L.S. Shapley. 1991 ISBN 0-7923-1016-0
8. J. Abdou and H. Keiding: *Effectivity Functions in Social Choice.* 1991 ISBN 0-7923-1147-7
9. H.J.M. Peters: *Axiomatic Bargaining Game Theory.* 1992 ISBN 0-7923-1873-0
10. D. Butnariu and E.P. Klement: *Triangular Norm-Based Measures and Games with Fuzzy Coalitions.* 1993 ISBN 0-7923-2369-6
11. R.P. Gilles and P.H.M. Ruys: *Imperfections and Behavior in Economic Organization.* 1994 ISBN 0-7923-9460-7
12. R.P. Gilles: *Economic Exchange and Social Organization.* The Edgeworthian Foundations of General Equilibrium Theory. 1996 ISBN 0-7923-4200-3
13. P.J.-J. Herings: *Static and Dynamic Aspects of General Disequilibrium Theory.* 1996 ISBN 0-7923-9813-0
14. F. van Dijk: *Social Ties and Economic Performance.* 1997 ISBN 0-7923-9836-X
15. W. Spanjers: *Hierarchically Structured Economies.* Models with Bilateral Exchange Institutions. 1997 ISBN 0-7923-4398-0
16. I. Curiel: *Cooperative Game Theory and Applications.* Cooperative Games Arising from Combinatorial Optimization Problems. 1997 ISBN 0-7923-4476-6
17. O.I. Larichev and H.M. Moshkovich: *Verbal Decision Analysis for Unstructured Problems.* 1997 ISBN 0-7923-4578-9
18. T. Parthasarathy, B. Dutta. J.A.M. Potters, T.E.S. Raghavan, D. Ray and A. Sen (eds.): *Game Theoretical Applications to Economics and Operations Research.* 1997 ISBN 0-7923-4712-9
19. A.M.A. Van Deemen: *Coalition Formation and Social Choice.* 1997 ISBN 0-7923-4750-1

20. M.O.L. Bacharach, L-A. Gérard-Varet. P. Mongin and H.S. Shin (eds.): *Epistemic Logic and the Theory of Games and Decisions*. 1997 ISBN 0-7923-4804-4
21. Z. Yang (eds.): *Computing Equilibria and Fixed Points*. 1999 ISBN 0-7923-8395-8
22. G. Owen: *Discrete Mathematics and Game Theory*. 1999 ISBN 0-7923-8511-X
23. F. Patrone, I. Garcia-Jurado and S. Tijs (eds.): *Game Practice*. Contributions from Applied Game Theory. 1999 ISBN 0-7923-8661-2
24. J. Suijs: *Cooperative Decision-Making under Risk*. 1999 ISBN 0-7923-8660-4
25. J. Rosenmüller: *Game Theory: Stochastics, Information, Strategies and Co-operation*. 2000 ISBN 0-7923-8673-6
26. J.M. Bilbao: *Cooperative Games on Combinatorial Structures*. 2000 ISBN 0-7923-7782-6
27. M. Slikker and A. van den Nouweland: *Social and Economic Networks in Cooperative Game Theory*. 2000 ISBN 0-7923-7226-3
28. K.J.M. Huisman: *Technology Investment: A Game Theoretic Real Options Approach*. 2001 ISBN 0-7923-7487-8
29. A. Perea: *Rationality in Extensive Form Games*. 2001 ISBN 0-7923-7540-8
30. V. Buskens: *Social Network and Trust*. 2002 ISBN 1-4020-7010-1
31. P. Borm and H. Peters (eds.): *Chapters in Game Theory*. In Honor of Stef Tijs. 2002 ISBN 1-4020-7063-2
32. H. Houba and W. Bolt: *Credible Threats in Negotiation*. A Game-theoretic Approach. 2002 ISBN 1-4020-7183-3
33. T. Hens and B. Pilgrim: *General Equilibrium Foundations of Finance: Structure of Incomplete Markets Models*. 2003 ISBN 1-4020-7337-2
34. B. Peleg and P. Sudhölter: *Introduction to the Theory of Cooperative Games*. 2003 ISBN 1-4020-7410-7
35. J. Thyssen: *Investment under uncertainty, Coalition spillars and marker Evolution in a Game Theoretic Perspective*. 2004 ISBN 1-4020-7877-3

KLUWER ACADEMIC PUBLISHERS – DORDRECHT / BOSTON / LONDON